Blood Brothers

BLOOD BROTHERS

Hiram and Hudson Maxim:
Pioneers of Modern Warfare

Iain McCallum

CHATHAM PUBLISHING

LONDON

This is the patent age of new inventions
For killing bodies and for saving souls
All propagated with the best intentions . . .
Timbuctoo travels, voyages to the Poles,
Are ways to benefit mankind, as true,
Perhaps, as shooting them at Waterloo
LORD BYRON: DON JUAN

Copyright © Iain McCallum 1999

First published in Great Britain in 1999 by Chatham Publishing,
61 Frith Street, London W1V 5TA

Chatham Publishing is an imprint of Gerald Duckworth & Co Ltd

British Library Cataloguing in Publication Data
A catalogue record for this book is available from the
British Library

ISBN 1 86176 096 5

Maps drawn by John Richards

Typeset by Dorwyn Ltd, Rowlands Castle, Hants

Printed and bound in Great Britain by the Cromwell Press, Trowbridge, Wilts

Contents

List of Illustrations   6

Foreword   7

1   Beginnings   13

2   Europe Beckons   39

3   The Years of Collaboration   68

4   Gas-bags and Flying Machines   95

5   The Parting of the Ways   122

6   Fame and Fortune   143

7   Marching as to War   168

8   Armageddon and After   192

The Maxim Family Tree   217

Select Bibliography   218

Index   220

List of Illustrations

Between pp64-65
Hiram Maxim aged 17. (From *My Life*)
Jane Budden. (Maxim Collection, Connecticut State Library)
Hiram aged 35. (From H P Maxim, *A Genius in the Family*)
Hiram's children. (Maxim Collection, Connecticut State Library)
The Maxim Electric Lighting System. (Maxim Collection, Connecticut State Library)
Maxim gun at Kensington 1885. (Vickers Archive)
Hudson Maxim 1881. (Hudson Maxim Papers)
Hiram's fight with Johnny Palmer. (Chatham Collection)

Between pp96-97
Prince of Wales firing Maxim gun. (Collection Dolf Goldsmith)
Rifle calibre Maxim gun. (Vickers Archive)
Hudson's pressure cooker. (Hudson Maxim Papers)
Hiram and Hudson Maxim 1888. (Hagley Museum and Library)
Chinese delegation and Maxim gun. (Vickers Archive)
Hiram and Sarah Maxim. (Collection of Val J Forgett Jr)
Jennie Morrow Maxim. (Hudson Maxim Papers)
Lilian Durban. (Lake Hopatcong Historical Society)

Between pp128-129
Hiram's flying machine in hangar. (Maxim Collection, Connecticut State Library)
The flying machine. (Vickers Archive)
Crashed flying machine 1894. (Science Museum/Science and Society Picture Library)
Hiram and Sarah at Baldwyn's Park. (Vickers Archive)
Hiram Percy Maxim *c*1900. (Maxim Collection, Connecticut State Library)
Hudson Maxim at home. (Hudson Maxim Papers)
Hudson and Lilian in car. (Lake Hopatcong Historical Society)
Maxim Park, Lake Hopatcong. (Lake Hopatcong Historical Society)
Hudson, Lilian and Dr Durban. (Hudson Maxim Papers)

Between pp160-161
Hiram and Maxim Joubert. (Science Museum/Science and Society Picture Library)
The Captive Flying Machine. (Philip Jarrett)
Boer Commando with Maxim guns. (Imperial War Museum: Q101770)
The Civilian Advisory Board, 1915. (Hudson Maxim Papers)
German machine gun section. (Imperial War Museum: Q23709)
Hiram at his drawing board. (Philip Jarrett)
Memorial to Hiram and Sarah Maxim. (Author)
Hudson Maxim's plaque. (Author)

Maps
The North-East United States   p12
South-east London and parts of Surrey and Kent in the 1890s.   p40

Foreword

During the last century an obscure New England family produced an inventor of genius and two other notable personalities, each of whom was to make his mark on technical developments which have shaped the modern world. The first and most remarkable was Hiram Stevens Maxim, originator of the automatic machine gun; the others were Hiram's brother the explosives scientist Hudson Maxim and Hiram's son Hiram Percy Maxim, pioneer of the gasoline engine and the horseless carriage. Because of the family connection, and since many of their inventions had to do with the business of war, they have frequently been confused in the public mind, each being credited with the achievements of the other two.

We are here concerned with Hiram Stevens and Hudson Maxim, who after a period spent working together were to carve out separate careers in England and the United States. In their day both men were honoured as celebrities. Hiram, described by the historian of the Vickers company as 'perhaps the most radically inventive engineer in [its] history, and certainly its most bizarre personality', will always be remembered for the eponymous weapon which wrought so much slaughter during the first half of this century. 'His name,' wrote Sir Basil Liddell Hart in his *History of the First World War*, 'is more deeply engraved on the real history of the World War than that of any other man. Emperors, statesmen and generals had the power to make war, but not to end it. Having created it, they found themselves helpless puppets in the grip of Hiram Maxim, who, by his machine gun, had paralysed the power of attack. All efforts to break the defensive grip of the machine gun were vain; they could only raise tombstones and triumphal arches.'

Hiram by no means confined his inventive talents to the Maxim gun, the fame of which has obscured his other achievements, notably in the field of explosives. In this last he found an eager pupil in Hudson, who was to make his own unique contribution. When the poet and playwright Tony Harrison set out to dramatise the contrast between the utopian promise of science in the years before 1914 and the perversion of science which in the event produced ever more sophisticated weapons of destruction, he chose as archetypal figures Fellow of the Royal Society Sir William Crookes, the German chemist

Fritz Haber and the American inventors Hiram and Hudson Maxim. His musical satire *Square Rounds* (complete with chorus of Munitionettes singing the TNT song) was premiered at the Olivier Theatre, London, in October 1992, when most of the audience would have heard of Hudson for the first time. The title of the piece refers to the legendary early machine gun built by James Puckle which was supposedly designed to fire round bullets against Christians and square bullets against non-believers, and during the play the Maxim gun, naturally, takes centre stage.

It was not the first occasion on which the Maxim gun had featured in the theatre. Hiram himself recognised the irony that his name would always be associated with a killing machine which he devised in his spare time rather than with his numerous other inventions. During a long and busy working life he registered hundreds of British and American patents, covering a broad spectrum of scientific endeavour from gas and electric lighting to chemistry, steam power, aviation, ordnance, and much else besides. Hiram was a compulsive, or, as he put it, 'chronic' inventor, with the ability to bring a very practical yet analytical intelligence to bear on whatever problem currently engaged his attention. At the same time, in an era of virtually unrestricted capitalism, he and his brother quickly developed an eye for the main chance and regularly, if not always wisely, committed themselves to the hurly-burly of the commercial marketplace.

Hudson went through life in the shadow of his older brother. Where Hiram led so the younger man followed, until, in what is surely a classic instance of sibling rivalry, they were eventually driven asunder. As to their association with the technology of mass destruction, the reader must decide in the light of the facts how far as individuals they should be held responsible or culpable, and how far they were simply responding to the pressures of the age. In recent years such attention as has been paid to Hiram has dealt mainly with the technical advances to which he made his contribution. His flying machine still rates a mention in connection with the earliest attempts at flight, while the development and tactical importance of the Maxim gun has been well covered by a succession of writers on military matters, nowhere more thoroughly than by Dolf L Goldsmith in his book *The Devil's Paintbrush: Sir Hiram Maxim's Gun*. By contrast Hudson's work as an explosives scientist has remained largely unrecognised.

Since his death Hiram has been the subject of just one full-length biography. In 1920 Paul Fleury Mottelay published *The Life and Work of Sir Hiram Maxim*, which was based on notes dictated by the inventor during the last year of his life and on articles in the technical and engineering journals. While this is useful as a general survey of Hiram's achievements (and includes, helpfully, a complete list of his patents), the author makes no attempt to come to terms with the personal battles in which he was involved or the ramifications of his private life, which could not in any case have been made public at that time. In consequence the book is tantalisingly incomplete and short on human interest. To an extent the gap is filled by Hiram Percy Maxim's memoir of his father, *A Genius in the Family: Sir Hiram Maxim through a small son's eyes*, which appeared in 1936, and by the excellent *Family Reunion, an incomplete account of the Maxim-Lee Family history*, privately printed in 1971 by Hiram's granddaughter Percy Maxim Lee and her husband John Glessner Lee.

Otherwise the researcher must rely on the contemporary record, the numerous books and articles written by and about the brothers and particularly the autobiographies which they published late in life. Unfortunately these are both deeply flawed, being primarily intended by their authors to pay off old scores, not least against one another. During their early years and for most of their lives, as Percy Maxim Lee reminds us, the two men 'competed for fame and fortune with passion, sometimes with bitter animosity, and always with flamboyant publicity. Both of them loved a rip-roaring fight; they were brilliant, ruthless and overflowing with virility and vitality, ambition and intellect; they were both egocentric and egotistical . . .' It is against this folk memory and this background that Hiram's *My Life* (1915) must be read. Surely one of the most self-serving self portraits ever written, a heady blend of conceit, exaggeration, self-justification and half-truths, it has done little to enhance his reputation. More sensibly, Hudson delegated the task of committing his thoughts to paper to the journalist and biographer Clifton Johnson, whose *Hudson Maxim: Reminiscences and Comments* appeared in 1924 and was reprinted after its subject's death with the title *The Rise of an American Inventor*. Of the two accounts, Hudson's is the more thoughtful, though it is also ridden with bombast and has, like Hiram's, to be taken with a large pinch of salt, since much that is important is omitted and much that is trivial and irrelevant is included.

Subsequent neglect of the brothers may be in part due to the secrecy that for many years shrouded their dealings with the military and naval authorities as well as with the Vickers company in England and the Du Pont company in the United States. There is also the difficulty that after Hiram died in 1916 his widow, Lady Sarah Maxim, appears to have destroyed many of the notes and records concerning his work which she kept and over which she had absolute control. It is possible that Hiram, who had a pathological distrust of lawyers and litigation, encouraged her to take this action; in any event only selected items eventually found their way to the Connecticut State Library at Hartford where they are now to be found in the State Archives. On the other hand Hudson's widow Lilian carefully preserved all her husband's letters and other memorabilia, which, after many years in the possession of the family, are now lodged in the New York Public Library. This collection, together with the papers relating to Hudson Maxim in the Du Pont archives at Wilmington, Delaware, gives a fairly complete picture of the man and his work, which makes it the more surprising that no biography has yet been forthcoming.

Unless otherwise indicated, quotations are taken from Hiram Maxim's *My Life* or from the personal papers of Hudson Maxim and his conversations as recorded by Clifton Johnson in *Reminiscences and Comments*.

My thanks are due to the directors and staff of the following institutions:

In England: the Public Record Office, Kew; the British Library at St Pancras and Newspaper Library, Colindale; the Patent Office, London; the Imperial War Museum; the Science Museum, Kensington; the National Army Museum, Chelsea; and Cambridge University Library, for access to the Vickers Archive and especially material relating to the Maxim Gun Company, the Maxim Nordenfelt Guns and Ammunition Company and Vickers, Sons and Maxim.

In the United States: the National Archives and Library of Congress, Washington DC; the Naval Historical Center, Washington Navy Yard; the New York Public Library, for access to the Hudson Maxim Papers; the Hagley Museum and Eleutherian Mills Historical Library at Wilmington, Delaware; the Connecticut State Library at Hartford, for access to the Maxim Collection; the Center for Maine History at Portland; and the Lake Hopatcong Historical Museum at Landing, New Jersey.

I am particularly indebted to David J Corrigan, Museum Curator, Connecticut State Library, and Martin and Laurie Kane of the Lake Hopatcong Historical Museum for help with illustrative material. Also to David K Brown, Philip Jarrett and Ken Skinner, who were kind enough to read the manuscript in draft and made valuable suggestions. Any errors that remain, whether of fact or judgement, are, of course, mine alone.

Acknowledgements are due to the following for permission to make brief quotations from published works: Percy Maxim Lee, *Family*

Reunion; Dolf L Goldsmith, *The Devil's Paintbrush, Sir Hiram Maxim's Gun*; HMSO, Charles Gibbs-Smith's *Aviation: an historical survey*; Cassells, Harald Penrose, *British Aviation, the Pioneer Years*; James E Hamilton, *The Chronic Inventor, the life and work of Hiram Stevens Maxim*; A P Watt, Literary Executors of the estate of H G Wells, for the short story *The Argonauts of the Air*; Faber and Faber: Tony Harrison's play *Square Rounds*.

Iain McCallum
Bath
November 1998

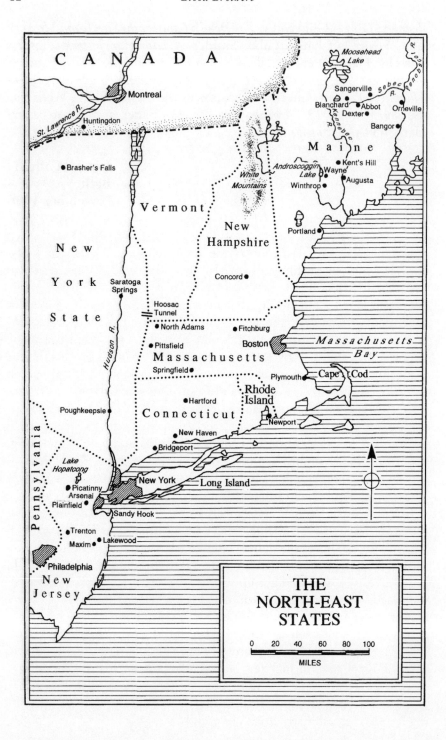

THE
NORTH-EAST
STATES

0 20 40 60 80 100

MILES

ONE

Beginnings

*The New England Yankees are the most ingenious and inventive of any
people in the world; and they excel because the Pilgrim Fathers, when
they landed on Plymouth Rock, bumped hard against necessity.*

HUDSON MAXIM

In the middle of the last century the state of Maine and especially its
northern region, pushing up towards the Moosehead Lake and the
Canadian border, was a harsh and unforgiving place. Winters were
long and hard, and while the hills and streams and woodland
abounded with wolves and bears and other wildlife, there were few
human inhabitants apart from the native Indians, who kept to themselves, and a number of scattered communities of European immigrants always on the move in search of virgin territory. Typical of this
enterprising breed were the forebears of Hiram and Hudson Maxim
(variously spelled Maxson, Maxie, Maxon and Maxham), descendants
of those French Huguenots who, according to family tradition, fled
from religious persecution to Canterbury in England before making
their way to Plymouth County, Massachusetts, where in Hiram's
words 'they could worship God according to the dictates of their own
conscience, and prevent others from doing the same.'

In due course Samuel and Saviah Rider Maxim struck out north-
wards to the district of Maine, where they settled near Winthrop and
Wayne on the shores of Androscoggin Lake within sight of the White
Mountains. Here they built a house and outbuildings, cleared the land,
raised livestock and brought up seven children of whom the youngest
was Isaac Weston Maxim, born in 1814.

Isaac, wrote Hudson of his father, 'was reared with an ax in his hand,
and got such education as he had in the chimney corner by the light of
a tallow candle.' Although barely literate, he was familiar with the
Bible and well read in such few books as he possessed, such as Parker's
Philosophy, *The Wonders of Nature and Providence* and Rollins' *Ancient
History* in several volumes, leather-bound. Despite his lack of schooling he was a 'great instructor to us children', being fond of verse and a
lively teller of stories, and his conversation 'was spiced with wit and

humour and irony'. He was also an ingenious and natural mechanic with a curiosity about the world around him and a creative flair which he was to pass on to his equally gifted sons. A pioneer in the classic mould, he was described by Hudson as 'of a dark complexion, with very fine straight hair, deep-set eyes, and an aquiline nose. His disposition was cheerful and he was philosophical about everything, yet he had spells of deep melancholy, probably induced by the struggle for a bare existence, which oppressed him heavily.'

Fortunately Isaac found a partner more than able to match his capacity for self-reliance and unrelenting hard work. Harriet Boston Stevens came of English stock, the oldest of a family of twelve raised by Levi and Hannah Stevens. Levi, known as Deacon or 'Old Brimstone' Stevens on account of his powerful convictions and fervent belief in hell-fire, had been born in Winthrop in 1787 and had subsequently moved north to Piscataquis County on the border of the wilderness. Here Isaac met and wooed Harriet, and when they were wed in 1838 they were said to be the best-looking couple that had ever stood up to be married in the town of Blanchard, Maine. Isaac settled down with his bride at Sangerville, a small township north of the Sebec river, where on 5 February 1840 Hiram Stevens Maxim was born, the first of eight children who were to arrive at regular intervals over the next seventeen years.

In later life Hiram and Hudson Maxim were frequently to acknowledge the debt they owed to their parents. Hudson in particular had a strong sense of family, believing that from Isaac the brothers derived their intellectual curiosity and inventive abilities, and from Harriet those qualities of character without which they could not have made their way in the world. From both Father and Mother, he wrote:

> . . . we inherited great physical strength and powers of endurance, but to our mother we mainly owe the unswervable will and courage to attack with ardent optimism any problem and face any circumstance or condition . . . She wasted no time on sentiment, [but] was a whirlwind of efficiency, and I can't remember a time when I didn't believe that she could do anything in the world possible of accomplishment by human hands . . . She was full of temperament and fire . . . and had what you might call hammer-and-tongs qualities. She was a thunderbolt . . .

Unsurprisingly these attributes were not always conducive to domestic harmony. During their early years the couple had to weather some stormy passages, not least when the community in which they lived was shaken by sectarian controversy. The United States, sustained by waves of immigrants each adding fresh revelations to the mix of religious experience, has always been a fertile ground for cults and dramatic conversions. Thus during the 1830s Isaac, brought up as a Congregationalist, was influenced by the teachings of William Miller, who argued from evidence in the Book of Daniel that Christ's Second Coming was imminent, and with it the end of the world. In due course it was announced that this would take place at midnight on 10 July 1843, at which time upwards of 50,000 New England Millerites or Second Adventists, including many of the good people of Sangerville, assembled with awe and no little trepidation to witness the great event. Not a few dressed in flowing robes in the belief that they were about to be taken up to heaven, but among the sceptics was Harriet, who, as befits the daughter of 'Old Brimstone' Stevens, held to more orthodox views and refused to go along with her husband.

The outcome, entering into family lore, was passed down to Hiram and in course of time Hudson, both of whom recorded it in their memoirs. As usual, Hudson's account is the more vivid:

> My father stood out in the front yard on that memorable watch-night and looked toward heaven. He stood there alone, for Mother didn't believe a word of the new creed. She jeered at him, and said, 'When you see Christ a-comin' call me, and I'll come out and catch hold of your coat-tails and go up with you.' The town clock struck twelve, and the minutes went slowly by as Father watched and waited. At last, despondent, crestfallen, sorely chagrined, he shed his ascension clothes and his new religion and went to bed.

Subsequently both Isaac and Harriet were to turn to spiritualism, but the effect of the Millerite experience was to instil in the Maxim boys an aggressive, free-thinking agnosticism and a distrust of organised religion which lasted all their lives.

Apart from such occasional differences Isaac and Harriet turned out to be well-suited. As was common enough in those days the newly-weds stayed at first with parents and other relations while they struggled to make a living from a succession of smallholdings. But as young Hiram was followed by Lucy (1842) and Henry (1844) it became increasingly difficult to make ends meet, especially when Grandpa and Grandma Maxim became infirm and had to be looked after. In 1846 Isaac gave up farming, sold his cattle and started a wood-turning business based on a workshop with two lathes at the nearby French's Mills. Drawing on an unlimited supply of timber from the surrounding forest, he set to work to make and sell wooden bowls, chairlegs, bedposts and carriage wheels. This enterprise proved more successful and a year or two later Isaac again improved his fortunes by moving to the village of Orneville, ten miles to the east, where a greater volume of water power enabled him to supplement his wood turning by running the community grist mill, grinding maize and other cereal crops for the local farmers.

For a time the family was relatively prosperous and two more children made their appearance, Leander in 1848 and Eliza in 1849. When he was eight Hiram went to the district school which opened its doors during the winter months, but his attendance was erratic since he was a loner by temperament, and he preferred to go his own way, doing odd jobs round the farms and helping his father in the mill. Most of all he enjoyed roaming through the forest and learning the secrets of fishing and trapping mink and musquash from the Penobscot Indians, with whom Isaac was on friendly terms. Quite early he discovered that he was strong for his age; in bouts of wrestling he could 'throw everyone except the big boys', and his determination and uncertain temper earned him a reputation for pugnacity. He also displayed a detached curiosity about the workings of nature unusual in one so young. Late in life he was to be reminded how he had taken a bottle-fly by both wings and slowly pulled them apart: 'This fly's wings,' he declared to his wondering friends, 'are not put in even; if they had been they would have pulled out at the same time.'

Like many a lad of his generation Hiram's early ambition was to become a sea captain, but as more mouths had to be fed with the arrival of Hudson (1853), Samuel (1855) and Frank (1857), his first responsibility was to contribute to the family income. By local custom Isaac had prior claim on the boy's earnings, and at fourteen he was apprenticed to a local coachbuilder aptly named Daniel Sweat, who got his money's worth out of the boy, insisting on his working all the hours in the day. Starting with wheelbarrows, Hiram soon graduated to the building of carts and wagons, for which he earned the princely sum of $4 a week paid in truck, that is, in goods obtained from the village store in lieu of cash. Despite this boost to the domestic budget the family lived for most of the time on the margin of survival. During the lean winter months the children were nearly always hungry, subsisting on limited rations of corn bread, beans and molasses with the occasional dish of fried pork. When the cow was in milk there was butter and cheese, the garden provided vegetables in season, and a favourite occupation was picking blackberries and raspberries in the surrounding woodland. Harriet kept turkeys and other poultry but eggs, which could be sold, were a rare treat.

Hudson Maxim first saw the light of day at Orneville on 3 February 1853 and was christened Isaac after his father. Throughout his youth he was known to the family as 'Ike', and only later did he repudiate his given name. Looking back from the halcyon years before 1914, Hudson was to marvel at the primitive conditions under which he and his family lived and at the progress in terms of civilisation that had since been made. 'I was born,' he wrote, 'halfway back to the Stone Age, for during the past seventy years at least one half of all the developments in the arts and sciences have taken place.' As infants he and his brothers and sisters were cradled in a big wooden bowl made by their father. The children were reared in one-story two-roomed shacks built of roughly hewn logs; they rarely washed and even in winter ran barefoot until well into their teens. Hudson recalled his mother making tallow candles and sometimes a rude lamp by means of a wick set in a dish of grease. 'An awful lot of bad-smelling smoke came from the lamp, but it served to light the house. That was exactly the light the cave men used.'

With Hudson little more than three years old the family made another of their many moves, this time back to the Sangerville district to live with Grandpa Stevens, and not long afterwards Isaac was offered the chance of a better situation at East Dover, where again he took charge of the gristmill and where he rented yet another ramshackle cabin. Meanwhile Hiram, now aged fifteen, had obtained a place as journeyman at a carriage manufacture and repair shop run by one Daniel Flynt in the nearby township of Abbot. Besides working and maintaining the water-powered machinery, the young man showed no small talent as an artist, which he put to good use decorating sleighs, carts and carriages with flowers and country scenes, making up his own colours out of red chalk, the juice of the alder plant and 'various other things that I could pick up'. For the next four years he travelled between Abbot and his home, spending six months at Flynt's workshop and six months helping his father in the mill, and it was during this period that he came up with his first invention.

'The grist mill at Abbot,' he wrote, 'supported a swarm of mice . . . While working in Flynt's carriage factory I used to make a few box traps in the noon hour and on Sundays, but the trouble with these traps was that when they had caught a mouse they were full and could not catch another until the first had been taken out. I therefore decided to make an automatic mousetrap, one that would wind up like a clock, and set itself a great number of times . . .' Other ideas he discussed with his father included a helicopter and a repeating gun mechanism, and he later claimed to have built the first tricycle in America, which featured wheel spokes held in place by tension rather than compression.

Soon Hiram was taking advantage of the time spent away from the family to make his mark with the local girls. Although uncouth in manner, he had a touch of wildness that appealed, and he was besides in Hudson's words 'a young man of wonderful personal appearance – hair jet-black, thick and curly, complexion pink and white; and, with his big brown eyes, beautiful teeth and fine physique, he was the centre of admiring eyes whenever he came back to the old home.' Hudson was naturally inclined to hero-worship his big brother, as indeed did all the family, while at the same time he was a little afraid of him and resented his cavalier ways. Hiram, he wrote, was 'rather an artistic genius, and had a passion for painting, but his eyes were perhaps a little too prominent and bold and staring, and he sometimes made me feel uncomfortable with his direct and rather glaring looks.' While everyone acknowledged Hiram's exceptional abilities, some re-acted against his insensitivity to the feelings of others and his some-what cruel sense of humour. All too typical was Hudson's disappointment when one day the two of them were out gathering berries. The younger boy caught three black crickets and 'Hiram said he would give me five cents apiece if I would swallow them. So I swallowed the three crickets, but I never got the fifteen cents.'

Nor, as time went by, could Hudson help noticing and resenting the partiality of their parents. Hiram, he observed, was the only child to be given a middle name:

> I think my father and mother took a somewhat greater pride in him than they did in the rest of us. They thought he was the great King Bee of the world, and used to hold him up to us as a model . . . Hiram was thirteen years older than I, and my youthful knowledge of him consisted mainly of glimpses. He was away working some-where or other all the time, actively engaged in the field of accom-plishment, and we younger members of the family looked up to him with marvelling.

In February 1860 Hiram, looking for wider worlds to conquer, persuaded Isaac to follow local custom by publishing a freedom notice in the *Piscataquis Observer* relinquishing all claims on his son's earn-ings. The young man's independence having thus been formally re-cognised, he struck out on his own, moving south to the comparatively large and prosperous township of Dexter, which boasted three wool-len mills employing English operatives from Lancashire. From here

Hiram travelled the district developing his skills as mechanic, wood turner and decorative painter. For the first time he had money to spend. Freed from the restraints of the family circle, he was able to pursue a lively interest in the opposite sex and to give free rein to his aggressive instincts, for he was quick to sense an affront and handy with his fists. It was not long before the ambitious artisan could see ahead of him enticing prospects for advancement, but then in April 1861 the guns opened fire at Fort Sumter, and his future and that of the whole Maxim family were thrown into uncertainty by the outbreak of the Civil War.

During the spontaneous outburst of patriotism that followed, Henry Maxim, aged eighteen, was among the first to volunteer for the army of the Union. Henry, like his older brother, was a born fighter, and after quarrels with Hiram that ended in fisticuffs he had already left home to work as apprentice carpenter with his uncle Amos Stevens at Upper Abbot. At first it was thought the war would be over in a few weeks and Henry, according to Hudson, 'imagined that he was going to have a sort of pleasure trip and see the world at the Government's expense'. He was soon disillusioned and turned against the war, writing in his letters home that he and his fellow soldiers were treated worse than the hogs back on the farm. Henry felt no enmity towards the Southern rebels and recounted how on the bloody battlefield of Antietam he took pity on a wounded Confederate soldier, propping him against a tree and giving him water. Seeing no purpose in the continuing carnage, he became depressed and took ill with chronic diarrhoea until eventually he was invalided out of the army. 'He left home strong and hardy,' remembered Hudson. 'When he returned, his teeth showed right through his cheeks, his face was so thin and drawn. He died in less than a month. We borrowed some clothes from the neighbours to wear at the funeral, and the neighbours lent us horses and wagons to drive to the graveyard and back.'

Henry's unhappy experience of the army in no way deterred young Leander Maxim, who like his brothers was a natural scrapper and who became so eager to join up that there was no holding him back. Calling at the depot, he was at first turned down on account of being under-age but then proceeded to impress the recruiting officer by successfully challenging all those present to lift, single-handed, a 400lb barrel of plaster. He was therefore posted to the 1st Maine Regiment of heavy artillery and in the spring of 1864, while serving in the Army of the Potomac, he was killed at Spotsylvania Courthouse in one of the series of engagements known as the Battle of the Wilderness. Many years later Hudson met one of Leander's old comrades who told him how it had happened. The regiment, having been issued with muskets, were fighting as infantry; they were also handicapped by a shortage of ammunition. As they stood shoulder to shoulder a rebel bullet struck Leander's brass cap badge of crossed cannon which presented a ready target to the enemy's sharpshooters. He fell without a word and his friend, still spattered with his blood, helped bury him next day. He was just sixteen.

Notwithstanding the feats of boxing and wrestling which feature so prominently in his account of these early days, Hiram had no wish become involved in the widening conflict. He had already decided that he needed to move away from Maine if he was to better himself in life, and so after a period training with the local militia in Dexter he went on his travels, leaving his brothers to join the colours. Afterwards his conscience troubled him, and he was fond of quoting a local medical man, Dr Springall, who assured him 'that it might be all right for those less gifted than myself to go to the war, but it was my duty to stay at home and work; also that I would find soldiering a very hard job indeed. So I made up my mind to give it up and refused to go on'. In any case some one had to look after the family. As he pointed out: 'My mother and my sister (Lucy) objected very strongly to my enlisting, and as it was the law that only two could be taken out of a family of three, I was exempt.'

For Hiram's activities during the war years there is no evidence other than the partial and incomplete account given in *My Life*:

> It has often been said that Maine is the best State in the Union to emigrate from, and I had long wished to get out of it, and go to some place where I could get more for my work, and have all my pay in money instead of partially out of the local stores. I had read a book about the St Lawrence River and Montreal, and I wished very much to see the great river. Accordingly I took the train one day and . . . found myself in Montreal – the first time in my life I had ever seen a large city.

From there he went to work at a threshing machine factory owned by cousins in upper New York State before going back to Huntingdon in Canada, where he practised his trade as decorative painter and saw 'some very lively times' working as a barman. There was at this time much interest in boxing following the widely publicised bare-knuckle fight between Heenan and Sayers, and Hiram seems frequently to have been involved in encounters with bullies or other difficult customers who were unwise enough to cross his path and duly received their come-uppance. 'During the two and a half years that I was roughing it in Northern New York and Canada,' he observed, 'I saw many fights and was in some of them myself . . . Fighting was purely a game of fisticuffs, no kicking being allowed, and what is more the man that got the worst of it harboured no grudge against the man who had beaten him at the game.'

Not until the summer of 1863 did he re-cross the border to stay with friends at Brasher's Falls in upper New York State. Earlier that year conscription had been introduced, but according to Hiram's version of events he was rejected in the ballot: 'I stood the draft, but did not draw a prize as my name did not come out'. Doubtless much relieved, and with anti-conscription riots raging in New York, Hiram was now in search of a regular occupation. Conscious of his lack of formal schooling, for notwithstanding his practical skills he was still only semi-literate, he spent the winter attending a writing school and reading such improving works as Ure's *Dictionary of Arts, Mines and Manufactures*. He was then glad to accept the offer of a job at $4 a day at his uncle Levi Stevens' engineering works at Fitchburg, Massachusetts. This was mainly engaged in the manufacture of automatic gas machines designed by Oliver P Drake, whose company was based in the nearby city of Boston. Lighting by gas had of course long been established but there was ample scope for development and the industry held out every expectation of a rewarding career.

At once Hiram set himself to master what was to him a fascinating new technology. 'These were glorious days,' he wrote. 'All my working hours were given to hard work and study. I left no stone unturned to become expert at everything I had to do.' As a result of his own observations and his skill in handling metalworking equipment the young man was able to suggest various improvements on Drake's original machine, which was designed to control and regulate the flow of gas for domestic and street lighting. He noticed that:

> . . . the gas made by the machines then in use was very rich at the beginning of the evening and inclined to smoke, whilst at the end of an evening it was thin and blue. I asked Drake if it would not be a good plan to make a machine that would turn out gas of a uniform density. 'Yes', he said, 'it would; that is the trouble with our machines . . . and there is absolutely no way to prevent it.' This set me thinking and experimenting, fully realising that carburetted air was much heavier than common air, and I made an apparatus to prove this.[1]

---

1. Mottelay, *The Life and Work of Sir Hiram Maxim*, p184.

It was not long before Hiram's resourcefulness and ingenuity came to the attention of the parent company in Boston. Within a year, and somewhat to the annoyance of his uncle Levi, he left Fitchburg to be appointed 'mechanical draughtsman' at head office with his salary doubled, and during the next four years he was engaged in supervising the installation of gas lighting systems in public buildings throughout New England. Despite severe competition for business, profits were good, and as his earnings rose so he spread his wings wider. Boston in the years following the Civil War was a place of hustle and bustle and there was no shortage of opportunity for anyone with ideas that might make money. Apart from his work for the Drake company, Hiram began to indulge his natural gift for invention, which manifested itself spontaneously as an insistent distraction from the work in hand. In 1866, after an abortive visit to the barber, he registered his first patent, for 'Improvements in irons for curling hair', and thereafter his mind was constantly preoccupied with fresh projects and enthusiasms. He was also stimulated in other directions. The big city offered many temptations to a young man on the loose, and the indications are that he sowed not a few wild oats before embarking on a more lasting affair with an English girl by the name of Louisa Jane Budden.

Brought up in London, Jane had emigrated at the age of sixteen to Boston, where she met Hiram a few years later. At first, repelled by his rough, provincial ways, she rebuffed his advances. 'It may surprise you to learn,' wrote their daughter Florence long afterwards, 'that when my father first began to pay attention to Mama, she was quite scornful. She was ashamed to be seen with him. He wore such awful clothes and had such terrible manners. He was constantly embarrassing her.'[2] Eventually, however, she succumbed to Hiram's ardour and persistence. With his dark good looks and piercing eyes he was hard to resist, especially as he gave every sign of being marked out for success. In 1867 the couple were married and set up home in Boston, and shortly afterwards Hiram's employers arranged for him to take a job as foreman and draughtsman with an associated concern, the Novelty Ironworks and Shipbuilding Company, situated on the East River in New York. Here it was that in September 1869 their first child, Hiram Percy, was born, in a rented house on Third Street, Brooklyn.

Hiram welcomed the opportunity to extend his experience into marine engineering, the Novelty company being mainly engaged in manufacturing reciprocating engines for the Pacific Mail steamship line. His main interest, however, still lay in gas regulating machines

---

2. Letter in Maxim Collection, Connecticut State Library.

and in particular a compact model devised by himself which, operating on a mixture of air and gasoline vapour, could be used to illuminate buildings located beyond the reach of city gas mains. While remaining in touch with Levi Stevens, he worked to such good effect that he was soon in a position to take out his own patents, on the basis of which he formed the Maxim Gas Machine Company with offices at 264 Broadway. In due course his machines, manufactured by a firm of locomotive builders in Paterson, New Jersey, were being used for the lighting of mills and factories as well as several of the A T Stewart group of hotels. They were also adapted as searchlights, some of which were supplied to the Imperial Russian navy, and as locomotive headlights, and for the first time Hiram began to reap the rewards of his independent initiative. Back in Maine, Isaac and Harriet and the rest of the family were pleased, though by no means surprised, to hear of this latest upturn in Hiram's fortunes, and young Hudson was delighted when in 1871 his brother suggested that he leave home to work for him in New York.

At a time when the average American could expect little more than two years of schooling before knuckling down to earn a living, Hiram and Hudson had fared relatively well thanks to their father's insistence on the virtue of self-improvement. Just the same neither was capable of writing or spelling to anything like an acceptable standard. Although Hiram continued to take time off to attend writing school, it is evident that he never did learn to put pen to paper with any degree of confidence. As for Hudson, he had also gone to school at eight, but so wayward was his attendance and so rudimentary the instruction he received that according to his own account he could not read properly until he was thirteen. He went to school for a further four years, but irregularly and only during the winter months because, he said, the family was too poor to support him and he had to pay his way by doing farm work and taking a job in a local brickyard at $25 a month. While these activities enabled him to rival his older brother in terms of physical strength when it came to boxing and wrestling, he was ill-equipped for any kind of career, and Hiram's summons to the big city was well timed.

'What,' wrote Hudson, 'a great, booming, buzzing confusion New York was to me on first acquaintance!' It was his first journey by train, and after arriving at the railroad station in his heavy country boots and checkered flannel shirt without a necktie, he rode in a horse car to the Fulton Street Ferry and crossed to Brooklyn, where Hiram lodged him in a nearby boarding house. Soon he was supervising the working of a gas machine used to light the evening performances of 'Howe's

Great London Circus and Sanger's English Menagerie of Trained Animals'. With the equipment mounted on a wagon 'as fancy as a fire-engine', Hudson travelled with the circus in and around New York, playing his part by riding in the parade dressed in ancient armour, plumed and helmeted. When not thus engaged he worked in Hiram's workshop polishing and varnishing and tidying up, between times taking in the sights, including Central Park, which he thought 'a somewhat marred piece of scenery, not as beautiful as I'd left behind in Maine'. On Hiram's advice he went to hear Henry Ward Beecher speak at Plymouth Church in Brooklyn, being duly inspired by the preacher's eloquence and his message that a man must seek to be positive in life, his reputation resting on what he does and not on what he doesn't do: 'How eagerly my ears and brain absorbed his words!'

Unfortunately there was not much that Hudson could do. As time went by he realised how poorly qualified he was to undertake any but the more humdrum jobs, and the little Hiram paid him scarcely covered his living expenses. After seven months he decided to quit New York, travelling home by sea by way of Portland, Maine. His first instinct was to consult his older sister Eliza, whom he regarded as 'the most human of all the members of the family. She had a big heart and a supreme sense of humour, and she was a good deal of a philosopher . . . Her common sense was wonderful . . .' Intelligent and with a mind of her own, Eliza had also insisted on going to school until she was fifteen and she now urged Hudson to 'get an education', which she saw as 'a sort of can-opener for the world'. Sadly neither of the Maxim girls was to survive beyond her twenties. Always delicate, Lucy took a husband who claimed to be a spiritualist preacher and doctor able to cure by the laying-on of hands, but despite his attentions she succumbed to consumption in 1870. As for Eliza, she was destined shortly to marry a farmer and to die, aged twenty-four, giving birth to their first child.

Following Eliza's advice, Hudson at first arranged to attend a private school at Winthrop, paying his way by doing chores for a family friend living nearby, and then in the autumn of 1871 he enrolled at the Maine Wesleyan Seminary at Kent's Hill. At the same time he decided to mark this new beginning by changing his name. He had long felt that Isaac, always shortened to 'Ike', was 'a damn bad name to be encumbered with', evidently because of its Jewish connotations, and despite his father's displeasure he made it known that henceforward he was to be called 'Hudson' after a lawyer friend whom he greatly admired.

At Kent's Hill, what he described as 'an institution chiefly devoted to making Methodist ministers of young men, and teachers of young women' offered Hudson courses in natural science and mathematics, literature and French, but to pay for his board and tuition he had to take on occasional labouring jobs besides teaching school (at $25 a month plus board) and working on the Boston and Maine railroad. This meant that he could only study part-time and for this reason, and because he was usually on the verge of destitution, his progress was slow. Fortunately from his point of view his parents decided in 1872 to make their final move, this time back to Isaac's home town of Wayne, seven miles from Kent's Hill, where they contrived to make a living from sewing and tailoring. Hudson's mother was therefore able to do the young student's washing and mending while he subsisted on a diet of baked beans, bread and milk.

For seven years Hudson struggled at Kent's Hill to qualify himself for some kind of worthwhile future, but he made little headway and could not help contrasting his situation with that of his older brother, who was going from strength to strength. In New York Hiram was impressing everyone with his abilities as a draughtsman, master of all trades and instinctive inventor. There seemed no end to his resourcefulness, gifted as he was with a lightning-like rapidity in separating the essentials of a problem from the non-essentials. He was now registering a steady and varied flow of inventions ranging from improved gas machines to an automatic sprinkler system for putting out fires (inspired by a blaze he had witnessed in Boston),[3] meters, pumps, dynamos and an apparatus for de-magnetising watches which remained in general use until the introduction of alternating current. His imagination was always alert to new possibilities. As he walked down Dey Street from Broadway one morning he noticed a powerful aroma of roasting coffee, and it occurred to him that here was a conspicuous waste of the essence of the coffee bean. This thought led in later years to a series of experiments designed to produce a coffee concentrate retaining all the flavour of the bean, a process which he patented in England but never succeeded in developing commercially.

By 1873 Hiram was senior partner in Maxim & Welch, Gas and Steam Engineers, proudly advertised as 'the business of Hiram S Maxim, inventor of the celebrated Maxim Steam Gas Machine, and A T Welch, late of the Gas Machine Co, Inventors and Manufacturers of Automatic Pumping Engines, Organ Blowers, Steam Pumps, Boiler Feeds etc.' It was a hectic life which left little time for his wife and children, and many years later Hiram Percy was to look back with

---

3. The system, activated by the fire, also rang an alarm at the fire station giving the exact location of the outbreak.

mixed feelings on the experience of growing up with his brilliant but
unpredictable parent. Clear in his memory was the family house in
Brooklyn with the cast-iron gate over which his father vaulted every
morning on the way to work, and his mingled delight and consternation at the elaborate practical jokes in which his father loved to
indulge.

Hiram, he thought, had 'blundered into fatherhood without giving
the matter any consideration', seeing it mainly as 'a means provided by
nature for perpetrating humorous misconceptions upon young and
inexperienced offspring'.[4] One escapade involved a petshop owner
who had jokingly promised Hiram Percy a puppy if he could pay for it
with a double-headed coin. At once Hiram repaired to his workshop,
where he made up a coin with heads on both sides with which the boy
duly confounded the petshop owner, who, however, refused to part
with the dog. For Hiram it was enough to enjoy the man's obvious
discomfiture, and much to Hiram Percy's disappointment he never did
get the puppy. Many of these pranks served an educational purpose,
one, for example, taking the form of an experiment with extremes of
heat and cold. The Maxims had a succession of cook housemaids
unkindly dubbed by Hiram 'Stupid One, Two, Three etc.' and to
illustrate the lesson for Hiram Percy's benefit he encased a poker in ice
before laying it on the shoulder of 'Stupid the Fifth', who, believing
that she had been burned, had hysterics and gave notice on the spot.

'My mother,' recorded Hiram Percy, 'had a flood of tears, the cook
packed up her belongings and departed in high dudgeon, and the
Sunday dinner was late and a very doleful affair.' Incidents such as this
afforded Hiram much amusement but were hard on Jane, who, conventional in taste and outlook, was temperamentally Hiram's opposite.
While she undoubtedly loved him, her concerns were almost wholly
domestic and she evinced only slight interest in the work which was to
him an obsession. Hiram, immersed in scientific and technical literature, had no time for music or the arts. They had little in common
apart from the daily round and an occasional visit to the theatre.
Throughout their married life Jane never ceased to be embarrassed by
his wayward brand of humour, and she tried constantly to persuade
him to behave in a more socially responsible manner.

It was no easy task. Not only was Hiram strong-willed and incorrigible, he was intolerant of opposition and quick to resort to his fists.
Hiram Percy described him as 'an extraordinarily powerful man and as
quick as a cat . . .' and recalled how on one occasion the two of them

---

4. This and subsequent quotations from H P Maxim, *A Genius in the Family*.

were approached in the street by a would-be mugger. 'Before the man had finished speaking my father grabbed him and actually boosted him up on top of the fence and pushed him over . . . [he] was the last man on earth to start monkeying with.' Hiram Percy also records that Jane did her best 'to soak some religion' into Hiram, occasionally persuading him to accompany her to church to hear Henry Ward Beecher preach wisdom and moderation, but 'unless the sermon was unusually interesting he would yawn so much, sigh so loud, squirm in his seat so continuously' that she had to give up the unequal struggle.

In 1873 the couple's first daughter, Florence, was born, and in the spring of 1875, when Jane again found herself pregnant, they decided to move out into the country, renting a house at Fanwood, near Plainfield, New Jersey. 'We had,' recalled Hiram Percy, 'a horse and carriage, a barn, a pig, some chickens, a cow, a garden, a blackberry patch and a hired man.' There were also two tame crows named by Hiram 'Moody' and 'Sanky', who flew down to be fed sitting on his shoulders. It was not, however, altogether a happy arrangement. Hiram, absorbed in building up his business and travelling ever more widely, spent most of his time away and came home only at weekends. As a result Jane found herself increasingly isolated and neglected, her situation being made no easier when in October a second daughter, Adelaide Louise ('Addie'), made her appearance. On the other hand the fact that she and her small son were thrown together meant that they developed a warm comradeship, and Hiram Percy was always to remember with affection how his mother would meet him from school and take him off for long, cosy chats at a nearby ice cream parlour.

Especially memorable were those rare occasions when the boy was able to enjoy his father's undivided attention. In 1876 a growing public interest in the march of technology led to the mounting of a Centennial International Exhibition at Philadelphia, the first such event to be held in the United States. All at once the Europeans were made aware of the swift strides being taken by American industry. As a German visitor observed, the relative shortage of manpower, an emphasis on practical education and a sound patent law were combining to 'bring machinery to the perfection at which we now see it, and to stimulate those great and small inventions which engender ever fresh undertakings . . .' Hiram, accompanied by his seven-year-old son, made the journey to Philadelphia and was enthralled and much influenced by the exhibits, which included a large Corliss steam engine, the latest model Gatling guns and a miniature steam railway, on which they took several rides. Many years later Hiram Percy was to recall his delight at the opportunity thus afforded to be on intimate terms with his elusive father. In the rooming house they slept in the same bed and at the exhibition, as they stood hand in hand before a 1400lb meteorite which Hiram said 'was not of our Earth', the boy felt 'a great wave of reverence and awe' sweep over him, an experience which was to inspire a life-long interest in cosmology.

In the same year Jane decided that the Fanwood home was in urgent need of redecoration. As usual Hiram was too busy to concern himself with such matters, and so it was arranged that his brother Frank Maxim, then in his twentieth year, would travel down from Maine to take on the job. Frank, the youngest member of the family, was according to Hudson 'tall, strong and wiry, and a very handsome fellow . . . a natural born mechanic', and after he had finished painting the house he went to help Hiram in New York. Within a few weeks the heat and humidity of the city took its effect. The young man was stricken with typhoid fever which was wrongly diagnosed so that he was sent to a 'smallpox pest house', where, though recovering from typhoid, he died of smallpox. This tragedy meant that of the eight members of the Maxim brood only three now survived, Hiram, Hudson and Samuel, who was without ambition and lived quietly with his parents on the homestead at Wayne.

Soon afterwards Hiram and Jane, tiring of the country life, moved back to New York, where they bought their first house, 325 Union Street, Brooklyn. Already Hiram was becoming engrossed with another topic increasingly debated in engineering circles, namely the use of electricity in place of gas for the lighting of public buildings. In 1878 a major international exhibition in Paris first drew the world's attention to the commercial possibilities opened up by the application of electricity to artificial illumination, and soon the scientific and popular journals could write of little else. Thomas Alva Edison, whose career in many respects parallels that of Hiram Maxim, was already blazing a trail which others were anxious to follow. Having also moved in 1870 to New York City where he perfected the printing telegraph, Edison had gone on to establish the first industrial laboratory at Menlo Park in New Jersey, developing the apparatus on which were based the phonograph and, in later years, the telephone and the motion picture. At this time his technicians were leading the way in the race to produce a serviceable, low-cost electric lighting system to compete with gas, and Hiram was one of many contenders in a strong field, prominent among whom were Charles Brush of Cleveland, Ohio, Paul Jablochkoff in France and the Englishman J W Swan.

Essential for success was the development of a long-lasting, low-voltage lamp to which it could be arranged for a regulator to supply, from some central source, a steady and unvarying flow of electricity. Working on the premises of the Union Metallic Cartridge Company at Bridgeport, Connecticut, Hiram experimented with dynamos, regulators and vacuum bulbs, and in September, 1878, he invented and patented the process of 'flashing' carbon filament lamps in a hydrocarbon vapour, only to find himself pipped at the post by Edison, who established priority for his own incandescent carbon filament lamp by a matter of days. This was a sad disappointment to Hiram, who ever after bore a grudge against Edison and continued to claim in his autobiography and in defiance of all evidence to the contrary that he was the originator of the electric light bulb. The experience, and the lengthy litigation that accompanied subsequent dealings between the lighting companies, implanted in him a virulent distrust of lawyers which was to stay with him all his days. Henceforward, too, he made every effort to ensure that each of his inventions was protected by the appropriate patent at the earliest possible stage.

Fortunately all was not lost. Hiram had already made the acquaintance of an enterprising New York businessman, Spencer D Schuyler, who was impressed by his record as an inventor and encouraged him to press ahead with developing and marketing his lighting apparatus in competition with Edison. The latter's Electric Light Company was incorporated in October 1878, with a capital of $300,000, and at about the same time Schuyler founded the rival U S Electric Lighting Company with more modest financial backing and himself as president. Hiram was appointed chief engineer with a generous salary, stock options and a free hand to pursue his researches. It was a major step forward in Hiram's career, and during the months that followed he justified Schuyler's faith in him by coming up with a system of arc lighting complete with its own dynamo and pressure regulator and a unique focussing device which enabled him to project a beam of light over long distances. He also devised and patented a process for producing pure phosphoric anhydride, a substance which absorbed the moisture in the pump used to create the necessary vacuum in his lamp bulbs.

Contracting the manufacture of equipment to the Bridgeport works, the Electric Lighting Company set about installing its machines and lamps in public buildings up and down the eastern seaboard of the United States, including the Grand Union Hotel at Saratoga Springs and in New York the Post Office, the Equitable Insurance Company headquarters and the Park Avenue Hotel. Hiram's curiosity was quickly aroused by the unexpected. At Saratoga Springs he noticed that the low musical note given off by the lamps was attracting

clouds of mosquitos, all male, and on investigation he concluded that
the note must be similar to the mating signal emitted by the female
mosquito. This oddity he described years later in a letter to the London *Times*:

> When the lamps were started in the beginning of the evening every
> male mosquito would at once turn in the direction of the lamp, and
> as it were face the music, and then fly off in the direction from which
> the sound proceeded . . . as the pitch of the note was almost identical
> with the buzzing of the female mosquito, the male took the music to
> be the buzzing of the female. I am neither a naturalist nor an
> entomologist; still I find much interest in this peculiar
> phenomenon.[5]

Soon Hiram was working so hard that Jane became concerned for
his health, urging him to exercise by going jogging with Hiram Percy
and encouraging him to relax by taking the children to Coney Island
to fish for blue crabs. She also tried to ensure that he dressed and
comported himself more in keeping with his new executive status, but
despite all her efforts Hiram kept up his reputation for absent-minded
eccentricity. Thus, always preoccupied with some intractable problem, he was notoriously careless with his belongings, which carried
stickers reading:

> This was lost by a damn fool named
> Hiram Stevens Maxim
> who lives at 325 Union Street, Brooklyn
> A suitable reward will be paid for its return

As the Lighting Company prospered so the couple began to lead a
more active social life. Hiram acquired a success symbol in the shape
of a 21-foot steam launch, the *Flirt*, which he moored at the Court
Street docks and on which he took friends and prospective customers
(though not Jane, who distrusted anything other than a ferry boat) on
cruises round New York bay. By 1880 the companies were competing
to come up with a reliable light bulb which would provide the kind of
soft illumination appropriate to domestic use. All the lamps so far
developed had been on the arc principle which gave a bright, harsh
light suitable only for public places, and as Hiram strove to adapt his
own version he was told by Schuyler: 'Maxim, light a house in New
York with these lamps and I'll sell your stock at 200 cents on the

---

5. *The Times*, 28 October 1901.

dollar.' This Hiram proceeded to do with such success that the competition, and particularly the group of businessman backing the Edison organisation, began to find themselves seriously inconvenienced.

What followed is still not entirely clear but Edward R Hewitt, an American who subsequently worked for Hiram in England, was later to reveal that early in 1881 the inventor was approached with an offer he could not refuse. According to Hewitt's account:

> The American companies engaged in exploiting electrical devices were constantly upset in their plans by being obliged to change their products before any money had been made from commercial exploitation. In desperation, they finally got together and made a joint deal with Maxim. The terms were extraordinary. Maxim was to go to Europe for ten years, at a salary of twenty thousand dollars a year, remain in close touch with all new [European] electrical inventions for the [American] companies, but under no condition was he to make any new electrical inventions himself during that time.[6]

On the face of it this arrangement, and the very large sum involved in paying Hiram off, is hard to credit: it reflects, however, a tactic which was and perhaps always has been a feature of the business scene, and it is wholly consistent with the dramatic events that were now to unfold.

News had already come from Europe that a second Electrical Exhibition on an even more ambitious scale was to take place in Paris in the autumn of 1881, and all the leading electrical concerns, including the U S Electric Lighting Company, prepared to show off their wares to the best possible advantage. In November 1880 the respected British periodical *Engineering* included an item on 'Maxim's New Electric Lamp: Mr Hiram S Maxim of New York has devised a focussing electric lamp . . . One good quality is the stoutness of its make, but there is no very novel feature in its mode of action.' Under the pressure of competition in the United States the company had in any case been considering extending its activities into Europe. The understanding reached by Hiram with the Edison syndicate was, therefore, accepted on condition that while fulfilling his side of the bargain he should act as the company's agent in Europe and complete the negotiations already in hand to amalgamate with a British firm, the Electric Light and Power Generating Company of Cannon Street, London.

Suddenly Hiram found himself a wealthy man, with the world at his feet. His first thought was to invest some of the money buying houses

6. E R Hewitt, *Those were the Days* (New York 1943), p113.

in Philadelphia,[7] and then in May 1881 his friend Nicholas de Kabath supervised the transportation of his apparatus across the Atlantic. As reported by *Engineering* in June:

> The incandescent system of electric lighting invented by Mr Hiram Maxim . . . has recently been introduced into London . . . and is now being exhibited every evening at the Albany Works, Euston Road. In point of mellowness and beauty it is one of the best incandescent lights yet made public . . . its most novel feature being the regulator by which Mr Maxim controls the supply of electricity to suit the number of lamps burning. The apparatus will shortly be removed to Paris for the forthcoming electrical exhibition . . .

On 14 August the inventor himself embarked on the SS *Germanic*. 'Eight days later,' he wrote, 'I arrived at Liverpool, and at ten o'clock that night was at the Charing Cross Hotel. I ate my first whitebait and saw the Thames for the first time. I was rather surprised to find how very small it was.' Next day he crossed to Paris. It was the most fateful journey of his life. Given the nature of his agreement with the Edison syndicate he knew that he was unlikely to be returning to America for some time. He could not have guessed that henceforward his contacts with his native land were to be limited to a few short visits at widely scattered intervals, but such indeed was to be the case.

Back in Maine Hudson, too, was once again on the move. Despite all his good resolutions and the years of part-time study, his academic record at Kent's Hill was far from distinguished. During the summer of 1878 he spent two weeks with Hiram and Jane in Brooklyn, where he was impressed by their growing prosperity and by a way of life which contrasted so sharply with his own. As usual he shared in Hiram's practical jokes, each according to Hiram Percy encouraging the other to more and more bizarre performances, and despite a bad bout of toothache he went over to New York every day with his brother, who, always pragmatic and inclined to be contemptuous of the value of academic education, was unequivocal in his advice: get out and make your own way, preferably westward, where so many opportunities were opening up for enterprising men.

Accordingly Hudson decided to leave the Seminary without graduating, defeated, as he subsequently maintained, by the handicap of poverty. He had acquired a smattering of physics, chemistry and French, took a romantic view of history and claimed to be inspired by

---

7. This turned out to be ill-advised, for the letting agents were unreliable and the houses had later to be sold at a loss.

the liberating philosophies of Darwin and Spencer, Thomas Huxley and John Stuart Mill. All this, however, was of little immediate practical value. For a time he taught school, taking a typically Maxim pride in 'making my pupils learn in spite of themselves . . . their progress was more rapid under me than any of their other teachers'. Alternative means of raising money included giving lectures on the fashionable topic of phrenology and devising from a distillation of the herb lobelia (a favourite of his mother's) in kerosene a specific for aching joints which he labelled:

MAXIM'S LIGHTNING CURE
Good for what ails you
For external use only

The vogue for patent medicines was then at its height, and doing the rounds of his customers Hudson was 'astonished at the almost universal satisfaction that the liniment gave . . . some told me they had been taking it internally with excellent results.'

Such stratagems were, however, short-lived, and in the autumn of 1879 Hudson decided to follow Hiram's advice and seek a more adventurous outlet for his frustrated ambitions. While at Kent's Hill he had befriended a classmate by the name of Alden W Knowles, who was 'about my age and equally poor . . . I found that he was exactly the kind of person I needed for a close associate because he would pick flaws in my ways and serve as a corrector of conduct . . . He was attracted by my powers of reasoning and we both felt a sense of intellectual companionship.' At the age of thirteen Knowles had been the victim of an accident during Fourth of July celebrations, when exploding gunpowder had shattered his leg. Since the injury could not be healed he was virtually bedridden for several years and he remained permanently lame. Soon the two young men were inseparable. Sharing their meagre resources, they talked and argued but never quarrelled since each valued the other's friendship too much to risk giving serious offence.

Among Hudson's personal papers there survives a scribbled note, probably his first essay in verse. Dated 1876 and addressed to his two closest friends, it reflects their aspirations for the future:

AWK and Stephen H
Receive my best regards
Perchance some day we may be great
And sung by future bards

As time went by Hudson and Knowles concluded that one way to achieve this happy outcome was to capitalise on the current enthusiasm for self-improvement on the part of a still largely illiterate population. The indications are that it was Knowles who led the way. Although not particularly literate himself, he had taken a course in penmanship and pen drawing, in which he proved to be quite a skilful practitioner. He therefore looked out a formula for making a strong and durable black ink, and the two friends decided to start a publishing and mail order business based on the sale of inks, steel pens and books of instruction.

Having agreed to follow the trail westward, the partners set their sights on the city of Columbus, Ohio, because of its central location, and after each had invested in a two-horse buggy they set out by different routes, reaching their destination after several months and peddling their wares along the way. For over a year Hudson and Knowles worked hard to establish themselves, achieving some success by wholesaling good quality pens and paper and capsules of coloured ink powder, but it was a difficult struggle in unfamiliar territory. Due to the damp climate both men suffered from bouts of malaria, and in the spring of 1881 they decided to return to the bracing air of New England. On their journey out they had passed through and been impressed by the beautiful town of Pittsfield, Massachusetts, high in the Berkshire Hills, and there they decided to relocate the business, using the money they had saved to produce what turned out to be a best seller, *The Real Penwork Self-Instructor in Penmanship*, advertised as 'The Greatest Means ever known for Learning to Write an Elegant Hand'.

Of this work, modestly sub-titled 'The Largest and most elegantly Illustrated Work on the subject of Penmanship ever published in the World', the partners had a dozen special copies printed on art paper, complete with frontispiece photographs of themselves. Being both enthusiastic admirers of Napoleon Bonaparte, regarded by Hudson as 'the greatest single human dynamo in the history of mankind', they were taken in the martial pose adopted by their hero in the famous portrait. The penwork book having sold a large number of copies by mail order or through canvassing agents, they went on to turn out a number of other popular works including 'Bible Pearls of Promise' and 'Golden Gems of Penmanship' together with a Family Record, an Autograph Album and the Lord's Prayer, 'attractively engrossed'.

As their enterprise flourished it seemed to Hudson and Knowles that they had succeeded in carving out a conventional if unexciting niche for themselves. Poor Knowles had constantly to battle against ill health, but Hudson was able to hold the business together while at the same time cultivating his social life and seizing every opportunity to practise his skills as a boxer. No doubt in emulation of his belligerent older brother, Hudson prided himself on his ability to deal with any physical challenge, and his memoirs are, like Hiram's, frequently punctuated by successful encounters with awkward customers who were foolish enough to cross his path. In fact, as Hiram prepared to depart for Europe, the younger man had good reason to believe that at last he could rest easy, free from the dominant influence which his brother had always exercised over him. So at least it seemed, but in the event appearances were to prove deceptive. Hiram was to continue to make his presence felt from across the sea, while already behind the scenes forces were at work which were to divide and scatter the surviving members of the Maxim clan.

In his wry and generous-spirited memoir, *A Genius in the Family*, based on recollections of his father, Hiram Percy Maxim records that when he was nine he travelled for the first time on his own from New York to Wayne to visit his grandparents. At the railroad station in Boston he was met by Mr and Mrs Haynes, friends of his father, who made a fuss of him and bought him a toy steamboat and put him on the train for Winthrop. For two months he stayed with the redoubtable Harriet, 'a very remarkable person', and Isaac, who was already in poor health and unable to do much walking, but with whom the boy had long talks about life and the world of nature.

Two years later, in July 1881, after Hudson's return from Ohio and shortly before Hiram's departure for Europe, the boy was taken by his father to what was to be the last family gathering at the homestead. As usual Samuel entertained the company with one of his celebrated readings, which made such an impact on young Hiram Percy that fifty years later he was able to recall the occasion as though it were yesterday:

> Before a high desk on a high stool would sit my uncle Sam . . . a small kerosene lamp furnished him light. The remainder of the room would be dark and mysterious . . . On the floor near the reader, his back resting against the wall, would sit my uncle Hudson. On this visit the book being read was Mark Twain's *Roughing It*. Uncle Sam had a wonderful voice . . . deep and resonant and dramatic. His wavy, jet-black hair, his flashing dark eyes and his remarkably handsome face suggested Wilkes Booth, the actor, my mother used to say. Hudson and my father were of the same type, and when a passage was read which impressed as humorous they would throw back their heads and laugh so loudly and savagely that it frightened me . . .

After this ten-day visit father and son left to return to Brooklyn: 'Little did I realise, when I said goodbye and drove away, that more than forty years were to pass before I should drive back over that road, and that I was never again to see my grandfather and my grandmother.'

The fact was that Hiram's marriage was already under severe strain. The nature of his work meant that he was often away from home, and during his travels he was tempted to seek out female company whether in the form of friendships or casual affairs. Early in 1878, not long after the family's move to Brooklyn, this indulgence took a more serious turn when he was captivated by a young girl by the name of Helen Leighton, with whom he got into conversation on a downtown street car. Pursuing her with his accustomed determination and undeterred by the difference in their ages (he was 38, Helen 15), Hiram overcame Helen's scruples by arranging for them to go through an apparent form of marriage, whereupon he established her as his mistress in rooms at Warren Street, Brooklyn. Here, as far as Helen was aware, they lived as man and wife, and in August she discovered that she was pregnant, a development which, although far from welcome, Hiram seems to have taken in his stride. In April 1879 Helen was delivered of a daughter whom they called Romaine, and Hiram contrived to maintain his double life, juggling the needs of his two families so that they remained in blissful ignorance of one another throughout this and the following year.

By good fortune the only one of Hiram's personal diaries to have survived[8] is for the year 1880, and this charts fairly clearly the course of his increasingly tangled private life. As is often the case with diaries, the entries start on a note of high endeavour (2 January: 'This day we commenced in dead earnest the experiments so long delayed on the electric light in a vacuum space . . .'), and they continue to describe work in progress at Hiram's laboratory in New York and the factory at Bridgeport. There are, however, frequent scrawled references to Helen, the 'Queen of Hearts', some explicit ('Un grand coup de Pussy'), one illustrated by a drawing of spermatazoa chasing ova, others thinly disguised in Morse code. And then, starting in the summer, there appear also short literary quotations, some in French, written in a different and obviously more educated hand.

Hiram had for some years been friendly with Mr and Mrs Charles Haynes of Boston and their daughter Sarah, and on 25 July, while visiting the city on a lighting assignment, he again called on them,

---

8. In the Maxim Collection, Connecticut State Library.

sketching next day in his diary a heart pierced with an arrow over the initials SH. From this time forward he took every opportunity to see Sarah, with whom he quickly developed a close and cordial relationship. Then in her mid-twenties, Sarah attracted him as much by her middle-class sophistication as by her slender figure and blonde good looks. Not only was she well read and able to speak French after a fashion, she was even teaching herself to master the intricacies of shorthand. For her part Sarah was flattered by the older man's attention and impressed by his energy and ambition. Unlike Jane Maxim she was happy to spend hours discussing with the inventor his latest ideas and enthusiasms, even offering to help by writing up his notes. Hiram's friendship with Sarah Haynes was as well known to Jane as to all the members of the family. Since it was based on a mutual interest in his work and appeared to be platonic in character, it was not at first seen as any kind of threat.

As the year 1880 drew to a close Hiram found it more and more difficult to reconcile these separate liaisons. Helen, to whom Hiram made no bones about his friendship with Sarah, became increasingly jealous and resentful of his visits to Boston. Early in the new year, moreover, her worst suspicions were borne out when a friend informed her that Hiram, whose activities were beginning to feature in the press, had for many years been married and already had three children. Making inquiries, she was soon able to establish that this was indeed so, whereupon she confronted her hapless partner, accusing him of taking advantage of her innocence and threatening to expose his infamy to the world at large. She also called to see Jane, who was, naturally, deeply upset to discover what her husband had been up to. Hiram, his back to the wall, responded by arguing that it was in nobody's interest for the affair to be made public. He urged Jane to overlook what he insisted was a temporary infatuation, and he proposed to Helen that he send her and baby Romaine a regular allowance on condition that she kept silent, went her separate way and in future called herself Nell Malcolm rather than Helen Maxim.

In the circumstances neither Jane nor Helen could do other than make the best of a bad business. Apart from the obvious consideration that she was dependent on her husband's earnings, Jane's main concern was to avoid scandal and protect her good name and that of her children. Helen had little alternative but to accept Hiram's offer of financial support and the conditions that went with it, especially when he broke the news to her that he was shortly going abroad. As for the villain of the piece, the whole sorry story provided another good reason why in the summer of 1881 he was glad to transfer his activities to Europe, far from the recriminations of both wife and mistress. As Hiram embarked on the *Germanic* he may have felt reasonably hopeful that the crisis had been resolved. This was, however, far from being the case. The episode remained as a skeleton in Hiram's cupboard which threatened at any time to emerge into the light of day, and many years later it was to cause him acute embarrassment at a critical stage in his career.

It was not long before there was sad news from Wayne. While in his fifties Isaac had suffered from an attack of measles which proved resistant to treatment and led to a slow but relentless decline. Eventually he had to take to his bed, and for four years Harriet nursed him with tireless and single-minded devotion. By the beginning of 1883 he was surviving only by means of a combination of morphine and whiskey and it was obvious that his days were numbered. In April Harriet wrote to Hudson to inform him that '. . . your father is failing fast his feet has begun to swell you no that is a bad sine I dont think he will live a month he wants to see you very much I wish you would come and see him before he dies whiskey is about all that keeps him alive from your mother H B Maxim he is about out of whiskey'. A few weeks later he was gone, worn down by years of anxiety and physical strain.

For all their admiration and respect for their father it appears that neither Hiram nor Hudson attended his deathbed or his funeral. Hiram had every excuse, being away on the other side of the Atlantic. As for Hudson, it is to be hoped that he did make the journey from Pittsfield, but there is no indication that he did so. Travel took time, and time was money. Despite this there can be no doubting the extent of Isaac's influence over his sons, who were frequently to assert that many of the ideas they were to develop had their origin with him. Hiram, like his mother, was not inclined to be sentimental and in his autobiography he makes little direct reference to Isaac, but it is reasonable to suppose that Hudson reflected the feeling of both men when he wrote long afterwards: 'My father meant a great deal to me. He was the source of an abounding inspiration, and as his genial, witty, philosophical and sympathetic face rises in my mind, I am stirred with kindly and affectionate memories.'

TWO

Europe Beckons

I was the first man in the world to make an automatic gun . . . which went into universal use throughout the whole civilised world. It is astonishing to note how quickly this invention put me on the very pinnacle of fame. Had it been anything else but a killing machine very little would have been said of it.

HIRAM MAXIM

When Hiram left America for Europe his relations with his wife had deteriorated to the point that they were scarcely on speaking terms. It was by no means unusual for successful men of affairs to be tempted into keeping a mistress on the side,[1] but nevertheless Jane found it hard to live with the knowledge that she had for so long been deceived. Nor was this her only concern. As time went on she found herself beset by more worries over the extent of her husband's attachment to Sarah Haynes. In fact even before departing Hiram had invited Sarah to join him in Paris in order to take up one of two secretarial posts created by the company to help him carry out his duties despite his inability to write grammatical English or speak more than a few words of French. The agreement with his backers in New York required him to supply them with translations into English of all the circulars, pamphlets and technical information put out by the competition, and for this task Sarah's language and shorthand skills made her eminently qualified.

Nor can Jane's misgivings about Sarah have been allayed by the first letter received from her husband, dated 7 September 1881 from his temporary lodgings at 86 rue Fauburg St Honore. Hiram addressed it not to her but to his twelve-year-old son Hiram Percy, and he wrote as he spoke in the direct style of the frontiersman, with scant regard for the niceties of syntax:

> I have been in Paris for about three weeks and as yet you have not written to me at all . . . I have sent for Miss Haynes she will be here

---

1. During the presidential campaign of 1884 the Democratic candidate Grover Cleveland was revealed as having earlier maintained both a mistress and an illegitimate child. This did not prevent him becoming president in 1885-89 and again in 1893-97.

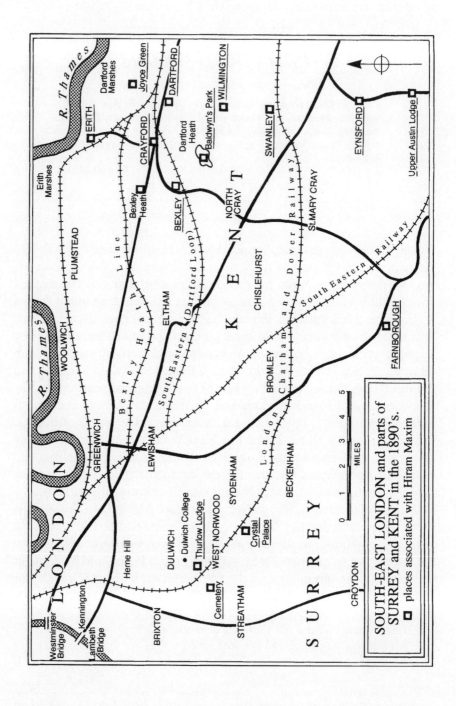

SOUTH-EAST LONDON and parts of SURREY and KENT in the 1890's.

☐ places associated with Hiram Maxim

on Sunday next if the ship comes in on time then she will help me but she will have no time to see much of Paris for a long time as she will have so much work to do . . . I will have her write to you all about Paris which I think is a humbug anyway. I am an American and see things like an American so I say Paris is all a Hollow show a complete sell. New York is the Finest city in the whole world . . . These Frenchmen are a going to have my Biography printed [presumably for the Exhibition brochure] to sell out here where people imagine I am a very great man . . . Miss Haynes is a going to write it . . .[2]

Clearly Hiram was feeling lonely and was missing his family, if not his wife. He had always been close to his only son, to whom he directed the few letters he found the energy to write home during the testing times that lay ahead. Although at first disposed to play down the attractions of the French capital, Hiram could not help being impressed as much by its cosmopolitan vigour as by the range and scale of the exhibition being assembled in the Palais de l'Industrie on the Champs Elysée. Ten years after the shock of defeat in the Franco-Prussian War, Paris was once again becoming the cultural and intellectual centre of Europe. As in the heyday of the Second Empire, the rich and famous could be seen spending lavishly in the fashionable theatres and restaurants, and in their wake thousands of British, American and German tourists were flocking to savour the delights of Montmartre and the Place Pigalle.

The International Electricity Exposition which opened its doors in September 1881 was a major event, described in one English journal as 'the most important and interesting scientific enterprise that the world has ever seen'. Even while Hiram was crossing the Atlantic the U S Electric Lighting Company had merged with the Weston Company of Newark, New Jersey, to exploit the joint Maxim and Weston patents, and so he found himself responsible for mounting the displays of both concerns. Throughout August the press reported on galleries being lit by the systems of Edison, Swan, Brush, Maxim, Weston, Crompton, Jablochkoff and others. The Electric Lighting Company of New York, it noted, was putting together a fine exhibit consisting of ten Maxim and two smaller Weston machines. Other attractions to catch the public imagination were an electric tramway and, even more startling, a telephonic device which enabled the performances of the nearby Opera House to be relayed to the exhibition halls: '. . . every note of the singers and orchestra, together with the applause of the

---

2. This and subsequent extracts from Hiram's letters from P M and J G Lee, *Family Reunion*.

audience and the sound of the steps of the corps de ballet, are transmitted . . . with perfect accuracy . . .'

Despite his lack of French, Hiram did his best with the help of Sarah and his technician colleagues to ensure that the Maxim-Weston display was as effective as possible. It was, however, evident that the competition was formidable, especially as Edison and Swan were well on the way to having the field to themselves. During the previous autumn Edison, having shipped a giant dynamo across the Atlantic, had supervised the construction of the first demonstration central power house at Holborn Viaduct in London, and in Manhattan his company was already planning to erect the first working power station to supply electric light to hotels, theatres and stores throughout New York. As for Swan, he had only a few months earlier created a stir by being the first man in England to install his electric lamps in a private residence, 'Cragside', the country mansion in Northumberland owned by his friend the armaments manufacturer Sir William Armstrong.

Throughout the period of the Exhibition the merits of the rival lighting systems were hotly debated. The reporter from *Engineering* commented favourably on 'the two powerful Maxim lights which are fixed over the principal entrance to the Palais de l'Industrie, each within a large optical projector by which its concentrated beams can be directed in any required direction', but from Hiram's point of view his overall verdict was disappointing. Although the Maxim display mounted 125 lamps, the illumination from them 'does not compare favourably with that of the other rooms . . . and is very far inferior to both the Swan and the Edison exhibits', the Maxim lamps having a 'smoky appearance due to a fine deposit of carbon forming on the interior surface of the globe'.

This and other indications are that despite the exaggerated claims Hiram was apt to make for his and the company's electrical apparatus, it could in reality be counted as no more than an also-ran. Having attracted 600,000 visitors the Exhibition closed in November with the French, who had provided nearly half the exhibitors, taking the lion's share of the prizes. Among the awards for the most successful individual contributions were Diplomas of Honour to the Americans Edison and Alexander Graham Bell. The U S Electric Lighting Company received no special recognition, Hiram having to be content with one of a fairly large number of gold medals presented to individual inventors. He was, however, consoled by receiving from President Grévy the Knight's Cross of the Legion of Honour, the first of several such decorations he was to accumulate in the years to come, and in which, like most of his contemporaries, he took great pride.

After the exhibition Hiram remained in Paris as representative of the New York lighting company. His new-found affluence enabled him to live in considerable style and he travelled widely on the Continent, supervising the sale and installation of the firm's equipment and occasionally crossing the Channel to negotiate with the Maxim-Weston associate company in London. Apart from sending money to his wife and family (as well as to Helen and Romaine) back in Brooklyn he made little effort to communicate with them, and in all probability they knew nothing of the terms of his agreement with the Edison consortium. The assumption therefore was that either Hiram would return to America or they would join him in Europe, but the situation was unclear and such letters as arrived from him only served to strengthen Jane's anxiety about the nature of his relationship with Sarah Haynes and the future of their marriage.

In February 1882 Hiram wrote again to his son, this time from 25 Avenue de l'Opéra, about visits he was about to make to London and Barcelona, after which, he assured him, he expected to return home. The letter, however, went on to make it plain where his real concerns lay:

> Sarah has spent the last three months at a French boarding school. She now speaks French very nicely. She could read and write it and speak plainly before she came but now she jabbers . . . You know of course that I am a Sir Knight now. Having been made so by the French president. I may be able to get another decoration in Spain . . . if I could take Sarah there and get her to try for it. A sweet and pretty woman who can speak French is a good thing to help in such matters especially if she is young and an American . . .

Throughout his stay in Paris, Hiram continued to grapple with the language, though in this as in other aspects of his affairs he relied increasingly on Sarah, who was now acting as his personal secretary and living, like him, in rooms at the Grand Hotel. She it may have been who suggested to Hiram that he scale down the payments to his former mistress on the grounds that after a year had passed she should be able to fend for herself. Outraged by this breach of the promises made to her, Helen turned to Hiram's uncle Amos Stevens, then resident in Philadelphia, who was moved by her plight and lent her the money to travel to Europe. Once there, Helen created violent scenes both in Paris and London, threatening to sue Hiram and expose him publicly, while he and Sarah retaliated by accusing her of trying to blackmail him. Eventually a compromise was reached whereby Hiram undertook to pay Helen a lump sum by way of settlement together with her return fare to America. It was also agreed that Hiram would arrange for friends of his to adopt young Romaine in order to ensure that she received a proper education.

Hiram had already discovered that in other respects Paris could be a dangerous place for the unwary. At first he naturally sought the companionship and advice of English-speaking people who were familiar with the city and its customs. One such was a Captain Graystone, of New Orleans, who struck up an acquaintance with Hiram before contriving, with the help of an equally plausible Irish accomplice, to steal from him a large sum of money. This was an especially chastening experience for someone as streetwise as Hiram, who reckoned to be able to look after himself. Afterwards he learned that 'Captain Graystone was really a notorious thief named Jack Hamilton, who had formerly been a prize-fighter and had seconded Heenan in the great international contest with Sayers . . . His accomplice . . . was none other than Johnny Palmer, another incorrigible scoundrel. This pair, it appears, had for years . . . carried on a lucrative business by similarly robbing Americans who came to Paris,' and Hiram swore to get even with them, however long it took.[3]

Despite these distractions Hiram had time enough to observe the European scene with a shrewd eye, noting the strong financial incentives that existed for business enterprise and surprised by the martial spirit which everywhere prevailed. In particular the French people were still smarting at the humiliation of their defeat in the war with Prussia. Ever larger sums were being allocated to rearmament and there was much talk of revenge and of a crusade to recover the lost provinces of Alsace and Lorraine. Passing through Vienna, he met an American friend who told him: 'Hang your chemistry and electricity! If you want to make a pile of money, invent something that will enable these Europeans to cut each others' throats with greater facility', and this thought took root in his constantly active brain.

Like so many of his countrymen before and since, Hiram had long been accustomed to owning and handling guns. His father and brothers had followed with interest the rapid development of firearms during and after the Civil War, and new rifles, revolvers and cartridges featured regularly in the popular journals and mail order catalogues. Back in Maine hunting for rabbits and partridges and even, for the more adventurous, bears had been a way of life. One of Hudson's earliest recollections was being let down by Hiram, who promised to

---

3. Illustrated Interview, *The Strand Magazine* (August 1894).

send him a fowling-piece from Fitchburg as soon as he could pronounce and spell every word in his Second Reader. The eleven-year-old worked hard to achieve this, but Hiram failed to deliver because Uncle Levi said it would be foolish to put a gun in the hands of someone so young. Nevertheless, in their teens the boys usually had access to guns and accidents were not uncommon, Hudson on one occasion being peppered with buckshot by Sam in a youthful quarrel.

In 1896, while addressing the Royal United Services Institution in London on the inception of the automatic system of firearms, Hiram stated that the idea had first occurred to him in boyhood discussions with Isaac at home in Maine. Then aged sixteen, he had even prepared drawings and models which he sent to his Uncle Levi at Fitchburg, but Levi thought them impractical and nothing came of them. A few years later the young man had occasion to fire a Springfield army rifle and, impressed by its powerful recoil, he suggested to his father that this otherwise wasted energy 'would be amply sufficient to perform all the functions of loading and firing, so that if the cartridges were strung together in a belt, a machine gun might be made in which it would only be necessary to pull the trigger, when the recoil would feed the cartridges into position, close the breech, release the sear, extract the empty case, expel it from the arm, and bring the next loaded cartridge into position.'

Since that time Hiram had kept the idea at the back of his mind, deterred from following it up by his other concerns and by the limited demand for such machine guns as already existed. He was of course familiar with the weapons invented by his fellow countrymen Richard Gatling and Benjamin Hotchkiss, the career of the latter being of particular interest since he was then working in Paris. Having during the 1860s supplied shells and cartridges to the French army, Hotchkiss was asked if he could improve upon the primitive volley gun, the 'mitrailleuse', which had so signally failed during the Franco-Prussian War. He thereupon established a workshop at St Denis on the outskirts of Paris, producing first a bolt-action repeating rifle (the rights of which he sold to the Winchester Repeating Arms Company of New Haven, Connecticut) and then a 'revolving cannon' firing small explosive shells. By 1881 the Hotchkiss gun was increasingly in demand, especially for mounting on warships, and the inventor was arranging for it to be manufactured by subcontractors throughout the world.

Hiram, meanwhile, was becoming restless and searching for a fresh outlet for his inventive energies. Barred from pursuing his electrical researches by the terms of his agreement with Edison, and with no lack of funds to invest in some alternative activity, he turned again to the drawing board. First, however, he had to fulfil his obligations to his employers by seeking to reorganise the Maxim-Weston company in London, which had its offices at 47 Cannon Street and a works at nearby Bankside on the Thames. Following instructions from head-quarters Hiram travelled back and forth to England in an attempt to reorganise what struck him as a woefully backward and inefficient operation, but he met with little success. Always inclined to be high-handed, he was soon at odds with the resident British management (whom he suspected of trying to undermine the Maxim and Weston patents) and losing interest in the uphill administrative struggle.

At this time, even as new projects were taking shape in Hiram's mind, the opportunity arose to settle scores with the confidence tricks-ters who had got the better of him two years earlier. Travelling by rail from Paris to London, Hiram recognised the Irishman Johnny Palmer among a group of men on the station at Rouen, and after a desperate chase ending on the footboard of the moving train he overpowered him and handed him over to the French police. According to his version of what followed Palmer was sentenced to five years in the New Caledonia Copper Mines. As for Jack Hamilton, Hiram had to wait rather longer but eventually he 'had him tracked, watched and finally sent into penal servitude for a variety of offences. This peculiar business took a long time . . . but I ultimately attained my object in view, and that is always the aim of my life.'

This same tenacity Hiram now brought to bear on what was to become the most renowned of his many inventions. While in Paris he had made preliminary sketches of an automatic gun based on the action of the Winchester rifle and now, finding himself in London with time on his hands, he took the first steps to put theory into practice. Having obtained permission to use the workshop facilities at Bankside, he worked there for a while before moving to a rented room in Cannon Street. Here he took on a draughtsman to assist him with the working drawings, and in June 1883 he registered his first patent for a 'Mechanism for facilitating the action of magazine rifles and other fire-arms'. In July this was followed by another for a 'Machine or battery gun . . . in which the feeding, firing, extracting and ejecting devices are operated by the force developed by the recoil of the breech block . . .' and the infant Maxim gun was born.

Not surprisingly all this extramural activity irritated the directors of the Maxim-Weston company, who sought to discourage their Ameri-can colleague by putting up the rent of his office and workshop space and declining to pay their share of his salary. But by now Hiram had the bit between his teeth. First he took his drawings to the

Birmingham Small Arms Company, which according to him showed little interest because of 'prejudices in the trade'. He therefore resolved to go it alone, leasing premises with an underground warehouse at 57D Hatton Garden, on the corner of Clerkenwell Road, acquiring the (mainly American) lathes, planers and drill-presses needed to turn out finely-machined parts of high-quality steel, and starting work on the prototype model. First indications were not encouraging. The workmen he took on to assist him proved less than competent, he had difficulty in getting the right materials, and when he went to order gun barrels from the Henry Rifled Barrel company he was warned against trying to build another machine gun. 'Don't do it,' the superintendent told him. 'Thousands of men for many years have been working on guns; there are hundreds of failures every year . . . You don't stand a ghost of a chance in competition with regular gunmakers – stick by electricity.'

Hiram responded with characteristic energy. He was still having to commute between London and the Continent, where the Maxim incandescent lamps continued to be in demand, one of the larger French electrical concerns installing them in the footlights of the Paris theatres and for lighting part of the Avenue de l'Opéra. During his absences the inventor delegated the work at Hatton Garden to two hired mechanics, but he soon found that they could not be relied upon to interpret the drawings unsupervised. He therefore decided to transfer his base from Paris to London, and rented a house at 6 Lancaster Road, Dulwich, where he was joined by Sarah Haynes. Now more than ever he was in need of an efficient secretary to keep his affairs in order: already he had come to depend on Sarah for the writing of reports and letters, and in any case she was by this time almost certainly his mistress.

An obvious difficulty the American had to contend with was the fact that the machine gun was far from accepted in military circles. Although a variety of rapid-firing guns, nearly all originating in the United States, was already in existence, none had proved altogether satisfactory. In addition to the Ager 'coffee mill' and other hand-cranked guns, the American Civil War had produced the Gatling gun, the most successful in terms of worldwide sales. This featured a number of barrels set round a fixed axis, firing in turn when operated by a crank lever, and it remained in American service from 1866 to 1911, being also adopted by the Royal Navy after 1872. In 1879 William Gardner of Toledo, Ohio, came up with a twin-barrelled gun which was rejected by the American authorities but favourably received in England, while at the same time the Swede Thorsten Nordenfelt started manufacturing at Carlsvik, near Stockholm, a weapon developed by the engineer Helge Palmcranz, with barrels arranged in a row. Firing a heavier-calibre bullet, this, like the Hotchkiss, was found increasingly valuable by the world's navies for the defence of warships against torpedo boats.

It is often alleged that the British authorities were backward in recognising the potential of the machine gun, but the record shows that of all the major powers Great Britain was the most actively concerned to evaluate each new weapon as it appeared. Thus the Gatling, Gardner and Nordenfelt were each in turn subjected by the War Department to thorough trials, the Gardner being recommended in 1881 as most suitable for both services on account of its light weight and relatively simple mechanism. It is true that few guns were purchased other than by the navy, but this was due not so much to innate conservatism as to the consideration that since such weapons were constantly being improved they could easily be acquired as and when they were needed. As the Duke of Cambridge, Commander-in-Chief of the army, was to remark in 1884, while he was 'greatly impressed with the value of machine guns, and feels confident they will, ere long, be used generally in all armies, he does not think it advisable to buy any just yet. When we require them we can purchase the most recent patterns, and their manipulation can be learnt by intelligent men in a few hours.'[4]

It is also true that the machine gun was consistently and everywhere underrated by army officers in the field. The reasons for this are not far to seek. During the Franco-Prussian War the French 'mitrailleuse', or grape-shot firer, developed in great secrecy by the Belgian firm of Montigny, had proved sadly ineffective. Deployed in the manner of artillery pieces, their positions had at once been revealed by clouds of gunsmoke, and they had been outranged and destroyed by the Prussian field guns. Nor did the Gatling, Gardner and Nordenfelt guns inspire much more confidence. Being mechanical and hand-operated they were cumbersome affairs; they needed teams of men to work them and they were apt to malfunction at moments of crisis under the pressure of battlefield conditions. Apart from being expensive to acquire and maintain, machine guns (referred to as 'mitrailleuses' until the early 1890s) were from a tactical point of view something of a Cinderella, unable to replace conventional artillery and seen as awkward or superfluous when used in support of cavalry or infantry.

Not that such considerations played much part in Hiram's thinking, concerned as he was only with the technical challenge. The principle

4. War Office memorandum: 'Machine Guns for Field Service', November 1886.

on which he worked was a simple one, namely to utilise the kick of a service rifle, hitherto regarded as at best useless and at worst dangerous, by storing this energy in steel springs which in turn activated an automatic process of loading and firing. The force of the recoil operated the breech-loading mechanism, discharged the gun and ejected the used cartridges in a continuously repeated sequence. As it evolved, the weapon, which could fire at the rate of ten rounds a second, came to incorporate a single, water-cooled barrel and an ammunition belt which ran through the action, held folded in a slot on the side. Capable of being worked by only two men, it was virtually self-operating, for as long as the trigger was pulled it kept on firing and ejecting a stream of spent cartridge cases until the belt ran out. The single barrel could also, significantly from a commercial point of view, readily be adapted to suit small-arms ammunition of any type or calibre.

As defined by its inventor, the automatic machine gun is one in which all the functions of loading and firing are performed by energy derived solely from the burning propellant powder, and in achieving this the Maxim gun was indeed revolutionary, representing what has been described as 'altogether the most remarkably innovative engineering accomplishment in the history of firearms'.[5] Just the same many obstacles had to be overcome before it could perform with consistent efficiency. While Hiram was lucky in that his key ejection mechanism was made possible by the recent introduction of solid-drawn, centre fire cartridges, teething problems obliged him constantly to modify the ammunition used, the springs gave trouble and the single barrel was prone to bulge under the stress of prolonged firing. Nevertheless, in the autumn of 1884 the prototype version was ready, and the first demonstrations were arranged for the benefit of special guests in the basement workshop at Hatton Garden.

Soon the newspapers were taking an interest in the weapon and, as the inventor recorded proudly in *My Life*, 'when it was reported in the press that Hiram Maxim, the well-known American electrician in Hatton Garden, had made an automatic machine gun with a single barrel, using service cartridges, that would load and fire itself . . . over six hundred times in a minute, everyone thought it was too good to be true . . .' The next step was to find a means of capitalising on his success, for as Hiram had learned from experience profits were made not from inventions but from their commercial exploitation. His first thought was to approach the leading gun producers in the United

---

5. Dolf L Goldsmith, *The Devil's Paintbrush*, p333.

States, who showed a marked lack of interest. He therefore looked to the possibilities in England, where the only comparable achievement was that of Sir William Armstrong, who twenty-five years earlier had produced his revolutionary breech-loading system for artillery, only to encounter much official resistance to the notion of his exercising any kind of monopoly over the manufacture of the guns. As Armstrong had pointed out in 1865 to the Royal Commission on Patent Law, 'the mere conception of primary ideas in inventing is not a matter involving much labour, and it is not . . . a thing demanding a large reward. It is rather the subsequent labour which a man bestows in perfecting the invention, a thing which the patent laws at present scarcely recognise.'

Unfortunately the most obvious prospect, Armstrong's own Elswick Ordnance Company at Newcastle-on-Tyne, was already licencee for the rival Gatling gun, but Hiram did not have to wait long for an alternative backer. Extolling the virtues of his automatic system for guns to a gathering of city men, Hiram aroused the interest of Robert Symon, a wealthy entrepreneur with extensive business and property interests in Mexico. Symon, alert to potentially profitable ventures, was intrigued by the force of Hiram's personality as well as by his ideas for new weaponry, on certain of which the two men were to work together.[6] Having called at Hatton Garden to see the prototype gun in action, he introduced the inventor to Albert Vickers of the Sheffield steel firm, which was suffering as the result of a general industrial recession and seeking to increase its stake in armaments production. The banking house of Rothschild was then brought into the negotiations, and Albert Vickers and Symon indicated to Hiram their willingness to finance his gun provided that he was prepared to devote himself full-time to its manufacture.

While naturally delighted by this development, Hiram was uncomfortably aware that he could no longer put off coming to a decision with regard to the future of his family. Jane had now moved with Hiram Percy, Florence and Addie from Brooklyn to Hyde Park, near Boston, to be nearer her relations. The children were becoming resigned to their father's long absence, and Jane was increasingly doubtful whether she and her errant husband would ever again be united, either in Europe or America. It seemed to be only a matter of time before the estrangement and the separation would have formally to be recognised.

The immediate cause of Hiram's break with Jane was a card he received from Hiram Percy in July 1884 conveying the good news that

---

6. Though nothing seems to have come of their joint patent, filed in October 1884, for 'an apparatus for adjusting, pointing or training cannon'.

he had graduated from high school. Already the boy was revealing something of his father's talent for engineering and a week later he received a reply, ostensibly from Hiram but in Sarah's handwriting and in a style unmistakably hers:

> By a card which you sent me I see that you have graduated. This marks a very important epoch in your life. You have now had more schooling than I have had, not so many years but at a vastly better school than had any existence when I was a boy. The time has now arrived for you to commence the serious part of your life . . . You must take up and learn some profession, and learn it well.
>
> I think it would be well for you to come to England and go into my little factory where the very finest kind of work is done. If more schooling is necessary, you will be able to attend one of the finest colleges in England [that is, Dulwich College] as I have hired a house very near one.
>
> Your mother and sisters will join me in England very soon. In the meantime, I want you to pack up your traps as soon as possible and take a steamer coming direct to London . . . You can get the money from your mother and as soon as you arrive I will refund it to her. You can go to work in my factory until the winter term commences at the college. I have a large house . . . so hurry up and come as soon as possible and keep me company until your mother and sisters come . . . Your affectionate father.

This letter provided Jane with ample confirmation that Sarah was living with Hiram and very likely in more than a secretarial capacity. It also strengthened her in the conviction that, despite her husband's assurances, there was little real prospect that she and the family would join him in England. Clearly it was Hiram's intention to lure his son away to London before leaving her and his daughters in the lurch, and this she was determined he should not be allowed to do. None of Jane's letters to Hiram has survived, but it appears that at this time she delivered an ultimatum: either he make arrangements for the whole family to go over to London without further delay, or she, considering their marriage to be at an end, would start divorce proceedings. As for Hiram Percy, his sympathies lay entirely with his mother, to whom he was devoted. He was also aware that she and his sisters depended very largely on money sent by Hiram, and if for any reason this was withheld he would have to take responsibility for them.

In his reply Hiram Percy announced that he intended to complete his education in America and so was unable to comply with his father's instruction to travel to England. It would seem that the letter was a spirited one, indicating his disapproval of the way his mother and sisters were being treated, for a week or two later Hiram wrote again, this time in his own hand and idiom:

> My Dear Son Percy,
>   Your curt letter to me is at hand.
>   So you choose to disobey the commands of your Father do you ?
>   Well it is not such a soft snap as you might suppose to drop into a first class position; however you will find this out for your self later. Who ever advises you to take the course you have taken is doing you a great injustice.
>
>   Your Father Hiram S Maxim
>
> PS   This will be a good letter to keep and read when you are about 25 years old.

This missive, with its single, oblique reference to Jane, effectively marked the end of Hiram's seventeen-year marriage. Added to the humiliation of the Helen Leighton episode, Jane could no longer tolerate her husband's indifference towards her, and his obvious intimacy with Sarah was the last straw. Divorce proceedings were therefore put in hand, although the formalities took time owing to Hiram's continued absence abroad. As for Hiram Percy, he was left in no doubt that he had been renounced by his father. That autumn he enrolled in the School of Mechanical Arts at the prestigious Massachusetts Institute of Technology, graduating in 1886 when he was seventeen, the youngest student in his class.

Meanwhile in London the negotiations between Hiram and his backers bore fruit when on 5 November 1884, at 57D Hatton Garden, the Maxim Gun Company was launched with an initial capital of £50,000 in 2500 shares of £20 each to 'carry on the business of gun, rifle, firearm and machine gun manufacturers, and manufacturers of all kinds of artillery . . . made in accordance with Maxim's patent or in any other way whatsoever . . .'. Albert Vickers was its first chairman and the Rothschild and Vickers families were soon to own a large proportion of the shares. Robert Symon and another businessman, Brodrick Cloete,[7] completed the board. Hiram became managing director with a salary of £1,000 a year,[8] which, added to the retainer being paid by the American electrical consortium, was more than

---

7. W B Cloete (1851-1915), pronounced 'Clootie', was a South African and an Oxford graduate.
   In later years he was appointed High Commissioner for Natal.
8. At this time a well-paid manager earned £300-£400 and a workman's wage was £60-£80.

enough to maintain him in the lavish style to which he had now become accustomed. Measures were put in hand to increase the output of guns at the Hatton Garden factory, and directors and shareholders looked forward to reaping sure and certain rewards.

By way of celebrating his success Hiram set about the purchase of Thurlow Lodge at West Norwood, which, close by the campus of Dulwich College, had once been part of an estate owned by Lord Chancellor Thurlow during the reign of George III. Set in its own grounds, the property adjoined the railway station at Dulwich on the London, Chatham and Dover line, so giving ready access both to London and the Continent. Thurlow Lodge was to be Hiram's home for most of his working life, and it was here, as well as at Hatton Garden, that early demonstrations of the prototype Maxim gun took place, apparently 'without obtaining the leave of anyone, neighbours or authorities, and the unusual noise created a great sensation in the vicinity'.[9]

The weapon was first revealed to the public at an International Inventions Exhibition at Kensington in the spring of 1885, the *New York Times* reporting that it was 'the marvel of the town . . . the authorities at the War Office regard it as revolutionising the science of war'. The inventor, it observed, was also working on a heavier version of his gun 'which would destroy torpedo boats as a terrier destroys rats'. Shortly afterwards a sample gun was presented to the prestigious Institution of Mechanical Engineers, and Hiram used the occasion to draw attention to its advantages vis à vis the Gatling, Hotchkiss, Nordenfelt and Gardner guns:

> All four of these depend upon hand power for performing the various operations of loading, firing and extracting the empty shells. Three of them are worked by a crank, while the Nordenfelt gun is worked by means of a lever, like an ordinary pump. As considerable force is required for working either the crank or the lever, the gun has to be mounted on a very firm stand or base, in order that it may not be rendered unsteady by the motion given to the handle. This necessity precludes the possibility of turning these guns with any degree of freedom . . .

None of this, argued Hiram, affected the firing of the Maxim, which was therefore not only more reliable and more accurate in its operation but less likely than mechanical guns to jam or break down when in action.

---

9. Hiram's obituary notice in *The Engineer* (1 December 1916).

It was not long before the new weapon and its charismatic American inventor caught the imagination of the establishment of the day. There was always something of the 'Barnum and Bailey' about Hiram Maxim, and he delighted in showing off his brainchild for the benefit of visiting dignitaries. 'Society,' he recalled, 'ordained that a pilgrimage to Hatton Garden in order to fire the new gun was a thing that must be done, and more than 200,000 cartridges were expended for the amusement of the "smart set", many of whom were anxious to fire it themselves.' Mr Matthey, the dealer in precious metals, brought along the Duke of Cambridge, Commander-in-Chief of the army, who was one of the first to be photographed seated behind the gun, blazing away through the clouds of smoke produced by black powder cartridges. Soon other luminaries such as Edward, Prince of Wales, the Duke of Edinburgh and the Duke of Sutherland were beating a path to his door, and Hiram had cause to complain that 'so much of my time was taken up showing the gun to visitors that it became necessary for me to work at night and on Sundays'.

Although officially he was still supervising the lighting company's European operation, Hiram now spent much less time on the Continent. In any case the Edison and Swan United Electric Lighting Company, advertised as 'manufacturers of glow lamps and incandescence arc and sunlight electric lamps to suit all kinds of lighting for Mines, Hotels, Cottages, Palaces, Churches and Clubs', was in process of establishing a dominance in the commercial lighting business which others found it difficult to challenge. Hiram continued to keep abreast of developments, setting up his own concern, the Maxim Electrical and Engineering Company, in London's Pimlico, but this does not appear to have made any great mark. As Hiram's electrical patents lapsed so his reputation as a gun inventor grew, opening doors on all sides, arousing the interest of the authorities and introducing him to a wider social circle.

Later in 1885 Sir Garnet Wolseley, veteran of the Ashanti and Zulu wars, visited Hatton Garden before arranging for the inventor to put the Maxim through its paces at the Pirbright ranges. Having shortly before returned from his unsuccessful expedition to relieve General Gordon at Khartoum, Sir Garnet was thoughtful, remarking that the gun was 'really wonderful' and could not fail to open a new epoch in the annals of warfare. Invaluable for halting what he described as the 'mad rush of savages', it would also, he believed, prove useful in civilised warfare provided that some means could be found of reducing the gunsmoke which obscured the gunners' aim and exposed it to the enemy's artillery fire. Jocularly he suggested that since the Americans seemed able to turn their hand to anything it would not be long before someone produced a machine that would manufacture fully grown men and women. Hiram amused his visitors by declining to take on the job, declaring that even if he were successful 'there would still be some old fogeys so wedded to old ideas that they would continue the old, slow, laborious, expensive and painful process.'

Over the coming years Hiram worked with his able assistant and foreman, the mechanic Louis Silverman, to make the Maxim gun simpler and more reliable in operation, filing a steady stream of patents to safeguard each successive advance against imitation by others. As he was later to explain: 'No one had ever made an automatic gun before; the coast was clear. Consequently I was able to take out any number of master patents . . . and to get very broad claims'.[10] The component parts were redesigned so that they were interchangeable and easily replaced in case of failure. Although chambered for the regulation .45in Martini-Henry rifle cartridge, the weapon was, as noted, readily adaptable to ammunition of other calibres, and it was soon evident that a heavier projectile would be valuable for guns mounted on warships. During the bombardment of Alexandria in 1882 and the suppression of the nationalist revolt in Egypt the use by the naval brigade of Gatling and Nordenfelt guns had so impressed one of the heroes of that campaign, Lord Charles Beresford, that in the summer of 1883 he argued in a lecture to the Royal United Services Institution that large numbers of these guns should be supplied to the fleet for use on land and to counter the threat posed by French torpedo boats.

Accordingly Hiram had already set to work to produce a larger-calibre weapon, first a 1in and later a 1.5in or 37mm version, which fired Hotchkiss-type explosive shells or hardened steel projectiles capable of perforating an inch of armour plate at a hundred yards. In April 1885 Albert Vickers wrote excitedly from his London club to Edward, the firm's founder: 'My dear Father, We fired the new one inch Maxim Gun today and it is a most absolute success. I would not sell it out and out for less than a quarter of a million, and if the Govt. had any brain they would pay that price and keep it secret . . .'. More cautious, Edward scrawled on the letter: 'This must be too sanguine. This Peace loving Nation with such an administration that we have would never purchase at any price.'[11]

---

10. Address to the Royal United Services Institution, 1896. In this way Hiram felt he had 'effectually prevented a great number of lawsuits that would surely have taken place'.
11. Vickers Archive.

The old man was right to the extent that the Admiralty was at first slow to order the larger weapon and the War Office showed no inclination to adopt either it or the rifle calibre version. In March 1885 it was reported in the company's minute book that a Captain Aubrey Patford was proposing 'to organise a corps to work Maxim Guns', and it was agreed that he be lent four rifle calibre guns 'to be returned at the end of the Soudan campaign'. Nothing seems to have come of this; instead six Gardner guns were sent to accompany the expedition. In September the War Office reported that: 'It is possible we might have a "Maxim" machine gun soon, in which case that would be recommended', but no action followed. Nor, although one of the first steps taken by the gun company was to assign foreign patents, was there much demand from abroad. Not until May 1886 was a sizeable foreign order for Maxim guns with armoured carriages received from Mexico, a breakthrough which, doubtless resulting from Symon's connections in that country, had the effect of emboldening Hiram and his colleagues to consider establishing a larger factory outside London.

Despite his commitment to the gun company, Hiram could not be restrained from pursuing other interests, his patent applications around this time ranging from apparatus for carburetting gas, the utilization of magnetism or electro-magnetism for the separation of metals and an 'intermittent water discharge apparatus applicable to the washing out of drains and water closets'. His efforts were, however, mainly directed to refining and improving the working of his guns and to considering other projects of a military nature, and he embarked on a series of experiments relating to explosives and projectiles especially for use in naval warfare. At the Admiralty there were reservations about the Whitehead automobile torpedo, which, although seen as a potent weapon, remained unreliable and inaccurate, with an effective range of only a few hundred yards. When, therefore, Mr Bryce Douglas of John Elder's Shipbuilding Works in Glasgow, described by Hiram as 'a very clever and well-known Scotch engineer', came to him with the idea of 'a gun of very large bore for throwing aerial torpedoes', he took it up with alacrity.

Hiram was inclined to agree that it was unrealistic to expect a torpedo to manoeuvre successfully underwater where it was subject to so many adverse forces, and he expressed his willingness to collaborate with Bryce Douglas and the Nobel Works at Ardeer in Scotland on the design of a heavy gun capable of firing a projectile with a massive explosive charge up to 600 yards. The expectation was that this 'aerial torpedo', armed with a delayed action fuse, would strike the water short of the target vessel and explode with sufficient force to shatter its hull. In May 1885 Hiram patented a large projectile 'chiefly designed for destroying ships' which was not taken up by the naval authorities, who rejected it on safety grounds. Just the same the exercise was to prove invaluable in that it brought home to the inventor the potential of recently improved explosives when used as the propellant and bursting charge in artillery projectiles.

Hiram needed no reminding that the smoke and heavy fouling produced by gunpowder cartridges were preventing his automatic guns from achieving their full potential. Attempts to fit a baffle to disperse the smoke which impeded the gunners' aim and revealed their position having been unsuccessful, he turned his attention to improving the nature of the propellant itself. At this time bullets and shells were still sent on their way by a variety of small- or large-grained gunpowders which were only partially burned up in the explosion. The hunt was on for a compound which, being almost wholly consumed when fired, would act as a more effective propellant and produce less smoke. According to Hiram's account his researches into the matter were stimulated by General Sir Andrew Clarke RE, Surveyor-General of Fortifications, who summoned him one day to the War Office. The Germans, he said, had developed more efficient, slower-burning powders which by gradually increasing the pressure behind the projectile were imparting higher velocities to their guns. What was the secret of these new powders and how could they be improved?

The answer Hiram claims to have found by means of a microscopic examination which revealed that the German 'powders' consisted of moulded blocks perforated with holes: as they burned so the holes grew larger, more surface was exposed and the gases of combustion were released at a progressively faster rate. This effect he reproduced in his laboratory by increasing the size of the grains in a similar powder which he patented in June 1885 but lacked the resources to manufacture for himself. He therefore 'went to Glasgow and had it made at the Nobel Works by Mr MacRoberts, the works manager . . .'. As always Hiram's version of events must be treated with caution. In fact the American Thomas J Rodman of the Ordnance Department at West Point was the first to come up with a larger grained, slower burning propellant which had seen service during the Civil War and was subsequently developed by the German explosives companies[12] in the form of prismatic brown or 'cocoa' powders. By the mid-1880s a number of researchers were seeking to improve on

---

12. Notably Köln-Rottweiler, who according to Sir Andrew Clarke were asking £35,000 for the licence to manufacture their powder.

these powders by exploiting the dynamite-related technology pioneered by Alfred Nobel and based on the explosive power of nitroglycerine. It was to take several years of experimentation before Hiram could enter the race to produce a true smokeless propellant.

Meanwhile, he was to discover that inventing the machine gun was easier than supervising its production on any scale. Looking back on his experiences at this time, he gave his advice to those thinking of setting up a manufacturing business – don't! As managing director of the gun company he could not avoid taking on a host of administrative chores which acted as an irritant, inhibiting his natural instincts as an inventor. This caused him to appear ever more intransigent and irascible, which in turn gave rise to production difficulties for which he was inclined to blame the shortcomings of British workmen and British equipment. In 1886 the immediate problem was how to compete effectively with the Nordenfelt concern, which, backed by a healthy order book, was about to develop a ten-acre site at Erith in Kent for the manufacture of its semi-automatic guns. It was quickly evident that this could only be done by moving from the cramped facilities at Hatton Garden to a properly-designed factory equipped with modern machine tools.

By the beginning of 1887 Hiram had perfected the latest version of his rifle calibre Maxim which, dubbed 'the little white gun' on account of its oxydised barrel, was ready to go into production. The improved weapon, henceforward to become the world standard model, received favourable reports from all sides. Mr Pratt of the firm of Pratt and Whitney, makers of the Gardner gun, saw it fired and was moved to comment that he would not have thought its rapidity of action to be possible. 'I would not have believed it if Mr Whitney had told me', he is said to have declared to the inventor, 'no, I would not have believed it if my wife had told me. But now I have seen it done with my own eyes.' But while the efficiency of the weapon encouraged the company to speed up the search for new factory premises, the British army still hesitated to adopt it. Not even the commendations of Sir Garnet, soon Field-Marshal Lord, Wolseley, who had been quicker than most to recognise its potential (and so was described by Hiram as 'one of the cleverest and brightest military men that I ever met'), were able to overcome this inertia.

The result was that the company's sales consisted mainly of piecemeal orders from war departments all over the world wishing to acquire one or two guns for purposes of experiment and trial, while others were given away in an attempt to attract publicity. In the autumn of 1886 Hiram had been invited by the Duke of Sutherland to a house party at his country seat at Trentham in Staffordshire where he was introduced to Henry M Stanley, the celebrated African explorer. Seeking to duck out of the ritual of church on Sunday morning, the two men went for a stroll in the grounds, and Hiram could not resist asking his fellow countryman why fifteen years earlier he had not succeeded in persuading Livingstone to return to England. Stanley replied, according to Hiram, that Livingstone 'did not want to come; he was well satisfied where he was, and was having a thoroughly good time.' A few weeks later the explorer was presented with a Maxim gun complete with arrow-proof shield which more than proved its worth during his expedition to the Congo to relieve Emin Pasha, who was holding Egypt's equatorial province against the Dervishes.

During the course of 1887 Hiram sought to cut through the veil of official reserve by cultivating his acquaintance with General Sir Andrew Clarke at the War Office and thoughtfully arranging for him to replace Albert Vickers, nominally at least, as chairman of the gun company. Soon afterwards this tactic bore fruit when three guns were supplied to the army's proving grounds at Enfield for testing and subsequently purchased. Just the same it was evident that the company would have to direct its main promotional effort overseas, and in *My Life* Hiram describes how during that summer and autumn, accompanied by Albert Vickers and the trusty Louis Silverman, he 'went into competition with the machine guns of all European countries', putting the Maxim through its paces at official trials at Thun in Switzerland[13] and La Spezia in Italy as well as in France and Austria.

Before long the success of the Maxim was causing so much alarm to the competition that the Nordenfelt company's European agent Basil Zaharoff, then on the threshold of his extraordinary career, was moved to take a hand. It was the inauspicious start to a life-long association between Hiram and the man destined to become not only a close colleague but also chief salesman and broker to the European armaments industry. In July both Maxim and Nordenfelt entered guns for a trial conducted by the Austrian army in Vienna and attended in person by the Emperor Franz Josef. The Maxim clearly outperformed the other contenders, firing 13,500 rounds without serious mishap (a broken mainspring was quickly replaced) and earning a round of applause from the onlookers by spelling out the Emperor's initials on the target.

The verdict of the trial committee was unanimous: '. . . it can be asserted that of all the systems of mitrailleuses hitherto tried, the

---

13. Where, employing Mauser cartridges, its superiority was so manifest that Hiram remembered the event as 'probably the turning-point in my life'.

Maxim is the best adapted to the purpose for which it is intended'. Only once during the proceedings did the gun break down, when Hiram assumed the fault to be due to negligence on the part of one of his mechanics, for nothing could shake his conviction that English workmen were habitually drunk and incapable. In fact Zaharoff, as he later admitted, had bribed the mechanic to sabotage the weapon. Not only that, but he went to the Austrian War Office and made every effort to disparage the Maxim and its American inventor, at the same time endeavouring to mislead the Viennese press agency by planting reports that the Nordenfelt and not the Maxim had been the more effective.

But as the Maxim attracted increasingly favourable attention such crude tactics were of little avail. During tests conducted by the Italian navy the gun's durability was highlighted when it was left for three days in the sea before being taken out and fired without difficulty. Following up the trial in Vienna Hiram wrote to the Hungarian Prince Esterhazy calling his attention to 'the advantages which the Maxim gun has over all others when used as a cavalry gun, or a gun to be fired from a galloping carriage . . .' Subsequently the company received an order for 131 guns from the Austrian government, and this was followed by another from Italy, the first nation officially to adopt it. At the same time the Prince of Wales, on a visit to Berlin, spoke enthusiastically to Crown Prince Wilhelm about the astonishing new weapon. The Prince had been strongly influenced by his friend Lord Charles Beresford's enthusiasm for machine guns, and in July the Crown Prince telegraphed to Beresford: 'Emperor very much interested in your [*ie* the new Maxim] machine gun. Wishes my regiment to try it as soon as possible. Please order immediately our calibre . . .'[14]

Matters then took a predictable course, given that at this time business relations with the Germans were good and Germany was regarded as Britain's most likely ally in the event of war with France and/or Russia. Through the agency of Lord Rothschild a meeting was arranged in London between Hiram and Friedrich Krupp, head of the leading German armaments firm, and in the spring of 1888 the gun company signed an agreement giving Krupps an option on the manufacture of 37mm Maxim guns for the Imperial Navy. This was to have far reaching consequences. A few months later, the Crown Prince having acceded as Kaiser Wilhelm II, a machine gun trial was held at the Spandau range near Berlin. After all the competing weapons had

14. Geoffrey Bennett, *Charlie B: a biography of Admiral Lord Beresford* (London 1968), p142.

fired the Kaiser is said to have gone up to the rifle calibre Maxim, laid a gloved hand on the barrel and said: 'That is the gun; there is no other.' He then arranged to purchase, from his own pocket, a number of 11mm Mauser-calibre Maxim guns for supplying to each of his crack Guards regiments.

At home, too, the Maxim was achieving greater recognition. In December 1887 Hiram wrote to *The Times* again setting out its advantages over other machine guns, particularly its greater accuracy at long range due to its automatic rather than mechanical action. Soon afterwards Capt F G Stone RA, Instructor in Fortification at the Royal Military College, described its operation in lectures to the Royal Artillery Institution and the Aldershot Military Society, arguing that the weapon 'will effect such a revolution in civilised warfare as has not been known since the introduction of breech loading rifles'. The experience of the Franco-Prussian War, he maintained, was no longer relevant, for the Maxim was immeasurably superior to the French mitrailleuse, with its clumsy mechanism, slow rate of fire and range limited to little more than 500 yards. He had taken a deep interest in the development of this 'new and most perfect machine gun' since it first saw the light at the Inventions Exhibition more than three years previously, and he went so far as to recommend that each army corps be equipped with a total of seventy-six Maxims of different calibres.

At the same time an article in another widely-read periodical, *The Engineer*, reinforced this positive view of the Maxim, commenting that 'the promises which we recognised [earlier] . . . have been fulfilled, and the weapon has been adopted extensively.' This was not entirely true. It was to be many months before the sales drive launched by Hiram and the company at home and in continental Europe was to bear fruit. Nevertheless, the orders already in the pipeline and the interest being shown by Krupp were deemed sufficiently encouraging for the gun company to raise the stakes in the competition with Nordenfelts. Accordingly it was decided to put in hand the building of a state of the art factory in a former textile works, 'a lot of large empty buildings . . . very suitable for the purpose' on the bank of the river Cray, a few miles from Erith, to which the manufacture of Maxim guns was transferred by stages from Hatton Garden during the spring and summer of 1888.

Besides having far-reaching implications for the local community, the move to Crayford involved Hiram and Sarah in a temporary change of residence from Thurlow Lodge to 'Stonyhurst', a rented house with five acres at Bexley Heath situated near the new factory. It also coincided with another momentous event. Early in 1888 Jane

Maxim finally obtained the divorce for which she had petitioned some three years earlier, and in May Hiram and Sarah Haynes were married. It was not before time. Tongues had been wagging, and it came as a relief to friends and colleagues alike when the couple were able to put their relationship on a respectable footing. The wedding was a quiet one, and its date was always kept secret[15] in an attempt to avoid the imputation that for many years Sarah had been Hiram's mistress as well as his secretary. It made little practical difference to their situation, although they were of course delighted to be able to occupy and furnish the new house in Kent as man and wife. Sarah continued to look after Hiram's personal and business affairs, not only writing his letters in a flowing copperplate hand but frequently signing them in his name. According to the company minute book she now owned 300 shares in the enterprise, and before long she was acting as her husband's proxy at board meetings during his absences abroad.

Behind the scenes, too, developments were taking place with a view to enhancing the gun company's performance and profitability. Unfortunately, although Hiram remained officially in charge of its operations, his inventive talents were not matched by his ability to deal with the everyday problems of the factory floor. As managing director he was the inevitable choice, but paperwork and what would now be called industrial relations were not his strong suit, and it was increasingly evident that radical changes were called for if the factory at Crayford was to come up to expectations.

The initiative was taken by Albert Vickers, an entrepreneur of flair and vision who was looking to create a lucrative combination of interests in steel, shipping and armaments. By 1887 his company had decided to enter into competition with Armstrongs and lay down plant for the building of warships and naval guns, and Basil Zaharoff was giving vigorous support to the idea of an amalgamation between the Maxim and Nordenfelt gun companies. Commercially the proposal made perfect sense, and it may indeed have influenced decisions already taken to locate the Crayford and Erith works close by one another and conveniently near to the Thames, Woolwich Arsenal and the main railway line. The Vickers family were all in favour, as was the Nordenfelt board. Only Thorsten Nordenfelt showed reluctance, having to be persuaded by colleagues more aware of the potential advantages for all concerned. Zaharoff was quoted as saying of the sabotage episode in Vienna: 'Maxim was furious, but he forgave me', and it

---

15. It can, however, be worked out from entries in the gun company's minute book. Until late May Sarah is entered as 'Miss S Haynes', thereafter as 'Mrs S H Maxim'.

seems that he got on better with Hiram, who admired his swashbuck-
ling style, than with the somewhat dour Swede.

Underwritten by two of the leading financiers of the day, Ernest
Cassel and Lord Rothschild, a loan issue of £1,800,000 to finance the
new enterprise was quickly taken up and in July 1888 the Maxim
Nordenfelt Guns and Ammunition Company came into being. The
chairman was Lieutenant-General Sir Gerald Graham VC of the senior partner, Nordenfelts, and Hiram Maxim and Thorsten Nordenfelt
were appointed joint managing directors. The American had good
reason to be pleased, for, in addition to his enhanced status, he received a generous allocation of shares and a bonus of £2750 as a result
of the merger. This was just as well, for having fulfilled the terms of
his agreement with the American consortium, he was shortly to cease
receiving payment from them or from the lighting company in New
York (later to be taken over by Westinghouse Electric), and in any
case his various electrical patents had long since been allowed to lapse.

Hiram's energies were now increasingly taken up with researching
the new explosives and finding ways of overcoming the safety problems which stood in the way of their adoption. He experimented with
propellant powders with and without the dangerous nitroglycerine,
heavy gun barrels produced from single steel forgings, and a novel gas
check or obturator to prevent the explosive gases passing the driving
band of the shell. At the same time he was busy organising the new
factory in Kent, for while at Hatton Garden a rough production line
had sufficed to turn out guns on a piecemeal basis, at Crayford it was
important to achieve a higher output to justify the higher overheads.
From the beginning the company resolved to make use of American
machine tools and the transatlantic approach to mass production. Following the success of the Springfield and other manufactories in turning out standardised small arms for the Northern armies during the
Civil War, the lathes, milling machines and other tools produced in
New England were rightly seen as the most advanced in the world.
American methods had already been widely adopted in England, notably for the manufacture of rifles at the Royal Small Arms factory at
Enfield, and in order to compete it was evident that the Maxim enterprise would have to introduce similar working practices.

It was with these considerations in mind that Hiram turned to his
brother Hudson in Pittsfield, Massachusetts, requesting him to iden-
tify one or two qualified technicians who could be relied upon to help
the company establish the works at Crayford on the right lines. Some
time in the spring of 1888, according to Hudson's account, Hiram
wrote him a long letter setting out the position and 'asking me to find

him a superintendent who understood the American method of making guns by what is called the interchangeable plan – that is to say
where all the parts . . . are exactly alike and a gun may be assembled
from those parts with certainty that every piece will fit . . . I looked
around and finally visited the Winchester Repeating Arms Works',
which was located at Springfield, some forty miles away.

As always Hudson had been following his brother's career with
interest and no little envy, for in recent years his own fortunes had
taken a turn for the worse. During the early 1880s the Knowles-
Maxim publishing business was doing well enough; at one time its
assets comfortably exceeded its liabilities and the partners were able to
open a Canadian branch at St Catherines, near Niagara. At the same
time Hudson was greatly intrigued by press reports of the Maxim gun,
so much so that he was moved to try his own hand at inventing. Hiram
had earlier been working on an 'apparatus for cooking, digesting or
similar purposes', and this idea Hudson now followed up, registering
his first American patent for a pressure cooker capable of roasting and
baking by superheated steam. 'Sometimes,' he wrote, 'it worked very
well,' but the safety valve proved unreliable and one day the cooker
exploded. 'The kitchen windows were blown out, the door ripped off
its hinges and the stove demolished.' Unsurprisingly only a few sample
cookers were ever made, and in any case other more pressing matters
were soon engaging his attention.

Hudson was now in his thirties and developing intellectual preten-
sions of his own, and when in the summer of 1886 he met a 'very pert
and pretty' young school teacher, Jennie Morrow, at a local reading
circle meeting, he courted her by what can only be described as a
process of literary intimidation. According to his version of their conversation Jennie was inclined to belittle his educational achievement
and so he determined to cut her down to size:

> Have you read Herbert Spencer's *Education*, the most important
> writing on that subject ever published?', I asked. She admitted that
> she had not. I asked her if she had read Darwin's *Descent of Man*, or
> any of the works of John Stuart Mill, John Tyndall and Thomas
> Huxley, and had she read Draper's famous *History of the Intellectual
> Development of Europe*? She said she had not, and began to look
> worried.

As well she might. Notwithstanding Jennie's intellectual shortcomings and the disapproval of Alden Knowles, who thought he was making a serious mistake, Hudson was in love for the first time and pressed

am Maxim, aged 17. Taken
ile he was working for Daniel
nt at Abbot, Maine. (From
Life)

Jane Budden of Boston at
around the time of her marriage
to Hiram Maxim. (Maxim
Collection, Connecticut State
Library)

Right: Hiram aged about 35, senior partner in Maxim & Welch of Center Street, New York City. (From H P Maxim *A Genius in the Family*)

Below: Hiram's children (left to right) Hiram Percy, Adelaide and Florence, New York 1878. (Maxim Collection, Connecticut State Library)

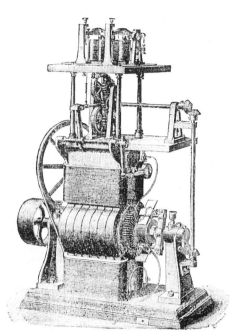

Régulateur électrique à courant

Régulateur

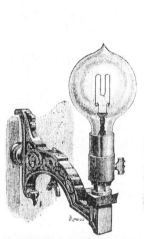

Lampe à incandescence

Double machine dynamo électrique en usage dans les laboratoires

The Maxim Electric Lighting System, as illustrated in the catalogue of the International Electricity Exhibition, Paris, September 1881. (Maxim Collection, Connecticut State Library)

Left: The Maxim gun on display in 1885 at the International Inventions Exhibition, Kensington, where it was awarded a Gold Medal. (Vickers Archive) *Right*: Hudson Maxim in Napoleonic pose, from the frontispiece of his and Alden Knowles' *Real Penwork and Self-Instructor in Penmanship*, published in 1881. (Hudson Maxim Papers)

Hiram grapples with the con-man Johnny Palmer on the footboard of the Rouen train as his companions 'endeavoured to reach me from the window and beat me off with sticks', from *The Strand Magazine* of August 1894. (Chatham Collection)

his suit until the young teacher agreed to accompany him to the altar. Their incompatibility was soon apparent. According to Hudson Jennie was flighty and neglected the housework, while as he put it: 'Our ideas and ideals were irreconcilable. Balls and parties and the social climb were her conception of successful achievement, while they were my abhorrence.' Over the next two years the couple's relationship was an uneasy one, nor was their situation helped by the steady decline of the Knowles-Maxim publishing enterprise. The introduction of the fountain pen and the typewriter hit the penmanship side of the business hard, and an attempt to diversify into printing proved a financial failure. By the end of 1887 the partners were losing money fast, having no option but to sell up most of their assets and settle with their creditors as best they could.

From Hudson's point of view, therefore, Hiram's summons from across the sea could hardly have been better timed. Unlike Knowles, who planned to go West and start another mail order enterprise specialising in silverware, he and Jennie were undecided about their future. Soon after receiving his brother's letter Hudson arranged to visit the Winchester factory, and after making discreet inquiries he was able to inform Hiram that William Griffin, one of the firm's most experienced foremen mechanics, was interested in joining the Maxim Gun Company, and that he in turn had identified a draughtsman willing to accompany him to England. Hiram, delighted, wrote by return asking Hudson to book a passage for himself and the two men, and in June the three of them crossed the Atlantic while Jennie remained in Pittsfield, doing her best to look after what remained of the publishing concern.

On arrival in England Hudson went to stay with Hiram and Sarah at their house in Bexley Heath, complete with carriage and groom. In the matter of the divorce his sympathies had always lain with Jane and young Hiram Percy, and it was no doubt with some embarrassment that he was now called upon to congratulate the newly-weds. Despite himself he was awed by the lavish scale of their lifestyle, in face of which he adopted the forthright, democratic stance of the American expatriate. Hiram, he observed, 'had acquired fame and wealth, and did a lot of hob-nobbing with the English aristocracy, but my instinct told me that his companions were no better than common folks. I had none of the false veneration so many have, and was not one who worshipped reputation or high position.' At the same time he was fascinated by the varied researches in which Hiram was engaged, and particularly the working of the famous machine gun. 'The Maxim gun,' he wrote, 'interested me greatly, and I studied it with all the energy that was in me. It wasn't long before I knew that gun by heart . . . and thus I became interested in explosives and ordnance, the field in which I have worked ever since.'

At Crayford Hiram showed Hudson proudly over the factory and the laboratory built specially for his use. The inventor was now at the height of his powers and engaged with so many projects that it was difficult for him to do more than keep a weather eye on any one of them. It was obvious that he needed all the help he could get, and, infected by his brother's enthusiasm, Hudson threw himself willingly into the role of assistant, studying Hiram's patents and starting a scrapbook in which he kept press cuttings on the subject of gun trials, military exercises and tattoos. During the months that followed he understudied Hiram in the laboratory and learned as much as he could about the work in hand. He received no salary, only expenses, but was issued with shares in the company by way of recognition of his services, which were to include writing the first manual of the Maxim gun. As for the two specialists from the Winchester company, the draughtsman quickly settled in to work while Griffin, 'a sharp-faced, clean-looking Yankee, about 45 years old', proved a great asset, at once taking charge of operations on the factory floor. His contribution, however, was short-lived since he disliked the English climate and after a few months decided to return home.

The period of Hudson Maxim's first stay in England, from June to December 1888, was one of signal importance for the brothers as for the budding armaments industry. Even as the Maxim-Nordenfelt merger was going through, scientists in Britain and on the Continent were coming forward with improved propellant powders which were soon to revolutionise the efficiency of small arms and artillery. In particular Hudson recalled that there was 'great excitement about a wonderful smokeless powder made and kept secret by the French government . . . I was given a few grains by a French employee of the gun company who had got it in Paris. I showed it to my brother and he was fairly beside himself with delight. We examined one of them, found out its composition and in two days we had duplicated and improved it.' This was the trail blazing 'Poudre Blanche Nouvelle' developed by Paul Vieille of the Ecole Polytechnique, which was first used in cartridges for the French small-bore Lebel magazine rifle and which, together with Alfred Nobel's 'ballistite' of 1887, is officially credited as establishing the modern propellant explosives industry.

As the British and German explosives companies drew together in an attempt to control the market for the new smokeless propellants, the directors of the Maxim Nordenfelt Guns and Ammunition

Company gave Hiram a free hand to conduct his experiments, and Hudson, helped by the grounding in chemistry he had received at Kent's Hill, proved an able pupil. In the years to come the two men were to share an active interest in this crucial field of research. In January 1888 Alfred Nobel filed his ballistite patent and by the year's end Hiram found himself in competition with two government scientists, Frederick Abel and James Dewar, who were working at Woolwich Arsenal on a propellant made to a similar formula but extruded in the form of strings or cords. This was, however, only one of many preoccupations, engrossed as he was with improvements to the machine gun, new designs of torpedoes and projectiles, a pneumatic gun for launching high explosive shells and, what was to become an increasingly pressing distraction, the design of a heavier-than-air flying machine.

Not all of these activities met with the approval of the company's directors. For Hiram's colleagues the most urgent priority was to market the Nordenfelt and Maxim guns which were coming off the production lines at Erith and Crayford. In this they were helped by the appointment of Lord Wolseley as Commander-in-Chief with responsibility for 'obtaining, holding and issuing to the Army all supplies and munitions of war'. Wolseley was widely regarded as a progressive by comparison with the elderly Duke of Cambridge, and it was probably at his instigation that in October 1888 the War Office at last placed an order for 120 rifle calibre Maxims, complete with accoutrements and ammunition, which were delivered over the next eighteen months.

This was an important step forward, which, together with the continuing demand from the Admiralty for the heavier Nordenfelt quick-firing guns, gave reasonable cause for optimism. There were, however, few signs of a continuing flow of domestic orders, so that to sustain even existing levels of production it was clearly necessary to stimulate a more vigorous demand from abroad. The redoubtable Basil Zaharoff, who had transferred his allegiance to the Maxim Nordenfelt company, was therefore given a roving brief, setting off initially on a tour of south and central America. Hiram's friend Nicholas de Kabath, now Russian consul at La Spezia, was recruited as the firm's agent in eastern Europe, based in St Petersburg. Hudson Maxim was appointed on a two-year contract as the company's representative in the United States. It was to him a welcome arrangement. He had never expected to remain in England and now he was glad to return home to Pittsfield, inspired with new ideas and ambitions and hope for the future.

THREE

The Years of Collaboration

*It is a common experience of inventors to discover that others have antici-
pated them. It is also a common experience for more than one inventor to
think of essentially the same thing and apply for a patent at practically the
same time.*

HUDSON MAXIM

At first expectations ran high for the Maxim Nordenfelt Gun Company, which according to its prospectus held 'the exceptional position
of covering the entire field for manufacturing automatic guns' and
started with 'a large and profitable trade in machine guns, quick firing
guns and ammunition'. As the first Maxims ordered by the War Office
became available they were supplied for demonstration purposes to
selected infantry units as well as to the School of Musketry at Hythe,
and in 1889 the company published the *Manual of the Maxim automatic
mitrailleuse, or rifle calibre machine gun*, drafted by Hudson Maxim
before his return to America.

The initial euphoria, however, was soon to evaporate as the demand
for guns from home and abroad turned out to be barely enough to
keep the wheels turning. At Erith and Crayford production difficulties
persisted, not least because the joint managing directors, former rivals
with strongly-held views, proved unable to work together. They were,
besides, distracted by outside interests. As well as spending long
periods abroad on sales and trouble-shooting expeditions, Hiram devoted much time to experiments with explosives or poring over plans
for a heavier-than-air flying machine, while Thorsten Nordenfelt was
concerned to develop the steam-driven submarines on which he had
been working for many years.[1] Never happy with the amalgamation,
he became increasingly embittered, uneasily aware that the growing
acceptance of the Maxim gun was steadily devaluing his own gun
patents. In January 1890 matters came to a head when Nordenfelt,
having acquired the patent rights to a new machine gun designed by a
Captain Bergman of the Swedish army, proposed to resign and depart

---

1. Between 1885 and 1890 four successive Nordenfelt submarines designed by the Reverend
   G W Garrett were tried out but none proved successful.

for Paris, whereupon the company brought a legal action by way of restraint, to prevent him re-entering the arms trade as a competitor.

Not until July 1894 were the prolonged and acrimonious proceedings finally settled in the company's favour, and meanwhile Hiram had to shoulder a greater share of responsibility for the day to day operation of the works at Crayford and Erith. Already, due to his overbearing style of management, the company had become embroiled in an industrial dispute of major proportions. A workaholic himself, the American had scant sympathy for requests from the labour unions for shorter hours. He had always, he was fond of saying, worked an eight-hour day – eight hours in the morning and eight hours in the afternoon – and he gave the impression that he expected others to do the same. The crunch came when the newly formed Amalgamated Society of Engineers, seeking better pay for a shorter working week, determined to take its stand on the issue of the introduction of piece-work. 'The struggle,' reported the journal *Engineering* in February 1890, 'was a long and costly one . . . fought out stubbornly on both sides . . .'. In the end 'the employers were victorious and were able for a time to impose conditions upon the men which were meant to utterly crush out the union', but a heavy price was paid in terms of resentment and low morale in the workplace.

In these circumstances no dividend was paid in 1890. The directors' annual report reveals that plant was not being fully utilised at Nordenfelt's Erith factory, which employed 700 men turning out quick-firing guns up to 6in calibre, while at Crayford, where Maxim guns were produced by a 400-strong workforce, strikes and layoffs all too often resulted in machines standing idle. This was in part because the large orders hoped for from the British authorities did not materialise. Up to 1892 only 169 Maxims were ordered by the Government, which, however, did negotiate for a licence to have them manufactured at the Royal Small Arms factory at Enfield in return for a royalty of £25 per gun. This meant that the works at Erith and Crayford found themselves even more dependent on demand from overseas. But while many countries were prepared to purchase small numbers of guns, with the Nordenfelt continuing to sell at least as well as the Maxim, this was insufficient to keep the factories in anything like full production.

Yet another disappointment was the failure of the company's representative in the United States to persuade the authorities there to adopt the Maxim gun in addition to the tried and tested Gatling. In 1887 three world standard rifle calibre Maxims had been despatched to the Washington Navy Yard, and in the summer of 1888 Brodrick

Cloete travelled to America to witness their testing at the army's firing range at Sandy Hook, south of New York. The result was that one gun was purchased 'for a more extended trial', and it was for this further trial that Hudson began to press directly after his return to Pittsfield in December.

He met with little success. Sentiment in America was isolationist and pacifist, and the government accorded low priority to expenditure on arms. Hudson did his best, offering to supply guns on loan, arranging for an American edition of his manual to be printed at Springfield, Massachusetts, and publishing details about the Maxim in an article in the *Scientific American*. It was mentioned that the weapon was 'soon to be tested at the [naval] proving grounds at Annapolis, with a view to procuring its adoption by the United States government', and in fact this took place in November 1889. Hudson supervised the proceedings, firing the test gun himself, but he had problems with the cartridges supplied and there were frequent stoppages. The Ordnance Board was 'impressed by the novel and ingenious features of the Maxim gun, and when its defects . . . are rectified . . . it may be a useful adjunct to our complement of machine guns'. Nevertheless, no more Maxims were purchased by the Board. It was to be many years before the US army showed any real interest in the Maxim, and in 1895 the US navy preferred to opt for the home-grown machine gun developed by John M Browning and manufactured by the Colt Firearms Company of Hartford, Connecticut.

In face of this official indifference Hudson turned his attention to what seemed likely to prove a more profitable line of business. As a result of his stay in England and his researches with Hiram into smokeless powders and high explosives, he was aware of the importance of the changes that had taken place during the 1880s, and in particular the implications of the ending of the centuries-old reign of gunpowder as the propellant and bursting charge for artillery shells. Since Alfred Nobel had invented dynamite by harnessing the destructive power of nitroglycerine, a new, science-based explosives industry had come into being dependent on dynamite-related technology. This was a development which in terms of its consequences may not unreasonably be compared with the advent of the atomic bomb. Since in the main the new explosives originated with and were marketed by the private sector, large fortunes stood to be won or lost and every step forward was accompanied by commercial rivalries and legal wrangling.

By 1889 all the leading European powers were experimenting with smokeless propellants which, displacing the old brown powders, could at once increase the range and rate of fire of artillery and small arms, clear the gunners' field of vision and enable their weapons to be more easily concealed. At the same time armies and navies were looking to introduce a new high explosive burster for their shells in place of gunpowder. Known to the French as melinite and to the British as lyddite,[2] this was many times more powerful than black powder, but serious objections were raised on safety grounds after trials revealed that shells filled with it were liable to detonate prematurely, destroying guns and killing or injuring the crew. The challenge was how to devise a means of propelling such shells without incurring this unacceptable risk. In 1887 Hiram filed a patent for an air gun 'chiefly designed for use with projectiles containing dynamite and similar explosive materials', and in the United States E L Zalinski, specialist in ordnance at the Massachusetts Institute of Technology, produced a gun which used compressed air to hurl a high-explosive shell up to a thousand yards. As it happens Zalinski had been financed by the New York businessman Spencer Schuyler, none other than Hiram's former chief at the electric lighting company, and this coincidence provided Hudson with his opportunity.

'I was,' he wrote, 'acquainted with Mr S D Schuyler, president of the Zalinski Dynamite Company, as he was one of the backers of my brother in his electrical business. Mr Schuyler had succeeded in making a sale of some [Zalinski] guns to the Government – one battery at Sandy Hook and another at San Francisco and also a small vessel with two guns mounted called the *Vesuvius*.' However, the air compressors were so clumsy and the range of the gun so limited that 'his organisation welcomed the suggestion made by me that a high explosive could be produced that would be safer, and a smokeless powder produced which could be safely fired from big guns and with greater range . . . which would thereby eliminate the air-compressors of the pneumatic system.' Already aware of Hiram's talents as an inventor, and impressed by Hudson's confidence that the brothers working together could come up with solutions, Schuyler indicated that finance might be forthcoming to establish a military explosives company in the United States.

Back in England Hiram was interested to hear from Hudson of this possibility, since he was trying to establish his own smokeless propellant, which he called 'Maximite', in the teeth of strong competition and difficulties with the British safety regulations. In January 1888, two years after the introduction of Vieille's BN powder, Alfred Nobel

---

2. Discovered in 1885 by the French chemist Eugène Turpin and based on picric acid (trinitrophenol).

had patented his ballistite based on a combination of nitrocellulose (or guncotton) and nitroglycerine. At the same time Abel and Dewar were working along similar lines at Woolwich Arsenal, collaborating on a propellant to which they gave the name 'cordite'. By the year's end Hiram had worked out a smokeless powder based on nitrocellulose alone or in combination with nitroglycerine, and, mindful of his failure to beat Edison to the punch in the matter of the incandescent lamp, he arranged to file three patents in quick succession, the last on 14 March 1889, more than two weeks ahead of Abel and Dewar. This did not, however, prevent the War Office from adopting cordite as the officially approved propellant for British service shells.

In this situation the Maxim Nordenfelt directors saw no reason to discourage their representative in America from seeking to develop the propellants and high explosives on which the brothers had both been working. Hudson had brought samples with him from England, and now he set out to find means of reproducing and if possible improving their composition. To this end he visited the American Xylonite explosives company at nearby North Adams, where he met Professor G M Mowbray, an old hand in the explosives business, who had achieved fame by manufacturing the commercial quantities of nitroglycerine used in the blasting of the celebrated Hoosac Tunnel. This, the first American rock tunnel, had been completed in 1875, linking Boston by rail with the mid-Western states. With Mowbray's help, Hudson was able to obtain the wherewithal to make nitroglycerine and nitrocellulose, and at his home in Pittsfield he proceeded to equip a laboratory to carry on his experiments, taking up with enthusiasm where he and Hiram had left off.

The times were auspicious, for it was rapidly dawning on war departments and particularly the naval authorities that smokeless powders were giving rise to what one expert was to describe as 'an entire reconstruction of our artillery'.[3] Already the British, French and other navies were conducting trials with cordite, ballistite and BN powder which, by progressively increasing the pressure behind the projectile as it travelled up the bore, imparted to it the higher muzzle velocities necessary to achieve longer ranges and pierce steel-faced armour plate. For the new propellants to be fully effective, long barrels were called for and these in turn accelerated the shift from muzzle to breech loading. Although the transition was the cause of much difficulty and confusion, by 1890 it was apparent that any army or navy failing to make use of smokeless powder was liable to find itself at a

3. J A Longridge, *The Artillery of the Future and the New Powders* (London 1891).

disadvantage. In the United States the Bureau of Ordnance requested the leading American explosives company, E I Du Pont de Nemours, to initiate research with a view to producing a modern propellant to replace the old prismatic brown powder. Alfred du Pont negotiated the purchase of a smokeless powder formula from the Belgian firm of Coopals, and the company set about establishing a manufacturing plant at Carney's Point, New Jersey.

But the Du Pont management had built their business mainly on explosives for blasting roads and railways, canals and tunnels. They had reservations about the commercial viability of smokeless propellant for military purposes, and the factory at Carney's Point was slow to deliver. Early in 1890, with the American army calling for the new powders and the navy contemplating firing trials to test them in projectiles for heavy guns, the market appeared to be wide open, and Hudson heard from Maxim Nordenfelt in London that the Winchester Arms Company was offering to buy the rights to make Hiram's powder in the United States. Reacting against any such idea, he proposed instead that Maxim Nordenfelt set up their own subsidiary operation in America to manufacture both smokeless powder and Maxim guns. At the same time he submitted samples of Maximite propellant for the consideration of Du Ponts and the Department of the Navy, pointing out that foreign royalties would not be payable for any powder produced in America.

At the Maxim Nordenfelt office in London Hudson's activities and especially his dealings with Du Ponts began to give cause for concern. The board took steps to safeguard Hiram's smokeless powder patents in the United States and elsewhere, and in August Hudson was summoned to London to report on the position in America. Having appointed his friend Schuyler as the company's temporary agent, he crossed to England on the liner *Majestic*. Interrogated by the full board, he was reassuring about the patent situation and argued persuasively for the setting up of a manufacturing subsidiary empowered to negotiate independently with the American authorities. Some directors were doubtful but others, including Ernest Cassel, were impressed by Hudson's ability, believing that he had the necessary drive and initiative to make a success of such an enterprise. An agreement was therefore drawn up authorising him to explore with Professor Mowbray the practicalities of establishing a plant in the United States to make smokeless powder. Hudson's salary was raised and Mowbray was paid $1000 a month together with a lump sum of $4000 for the use of his nitrocellulose and nitroglycerine patents.

Gratified by this satisfactory outcome, Hudson remained in England for several weeks renewing and extending his contacts there.

With Hiram he attended a function at the Army and Navy Club where the two of them, both zealous non-smokers, were repelled by having to endure the fumes which accompanied the ritual of after-dinner cigars. Throughout he stayed at Bexley Heath with Hiram and Sarah, in whose company he also made his first excursion to Paris, where he was impressed by their (and particularly Sarah's) command of French and dined with Basil Zaharoff, recently appointed as the gun company's foreign adviser, with whom Hiram was now on the friendliest terms.

Back in Pittsfield, Hudson set about laying the groundwork for the proposed new company, consulting Mowbray on the design of powder mills for the manufacture of explosives and smokeless powder. He wrote to the Ordnance Board requesting the loan of the Maxim gun at Sandy Hook, as 'I wish to get prices from American manufacturers for the building of [the gun] for the US Government', and he renewed his advances to the Navy Department with the object of persuading them to adopt the Maxim smokeless propellant. In August there was a further encouraging development when Du Ponts reported on tests carried out on Maximite at the Springfield Arsenal: 'The results are certainly good, the pressure and velocity increasingly regularly with increased charges . . .' On the other hand in London doubts were being raised about the notion of an American subsidiary. Hiram continued to be supportive but others, worried by the general situation of the Maxim Nordenfelt company, were opposed to taking on additional financial commitments and concerned about a possible loss of patent rights.

In October 1890 Brodrick Cloete travelled again to New York where he met Hudson at his hotel and conveyed to him the board's reservations about the proposed American enterprise. Next day he wrote by way of confirmation to advise caution: 'On thinking over all you said yesterday, I cannot refrain from stating that I think you ought to be careful not to attempt to move in so serious a matter as the formation of any Company . . . which cannot but militate against our own interests in the future.' This was something of a blow to the expectations of Hudson, Mowbray and Schuyler, and from North Adams Mowbray wrote to Hiram pointing out that it would be easier to manufacture smokeless powder in America than in England because raw materials were cheap and 'we are not trammelled with licences and Inspectors and numerous laws relating to explosives'. The indications were that a suitable plant with access to rail and sea could be built for some $10,000, while, 'as our product can be shipped immersed in water', it could be despatched with safety to any destination.

Hiram now decided that the time had come to make his first visit to America since his departure nine years earlier. Carrying with him 'a large number of specimens of smokeless powders in a leathern trunk', he announced to reporters in New York that because in England 'the laws are too stringent to permit of the erection of a powder factory', he had come to explore with his brother the possibilities for establishing a suitable works in the United States.[4]

Nothing, however, came of this initiative. Hiram was put out by Hudson's talk of an improved smokeless powder which he claimed to be making to his own formula, and the gun company remained unconvinced. Its financial problems were no nearer solution, and when Hudson's two-year contract came to an end it was not renewed. Nevertheless early in 1891 the company arranged for MacRoberts of Nobel to assess the quality of his smokeless powder and of Mowbray's guncotton. In April Hudson was again called to London, where he took a more independent line, giving the board an option of six months to approve his organising an American subsidiary to develop what he described as 'his inventions'. Further payments were made to him 'provided he transfers all patents re guncotton and smokeless powder to the Company', but as the months went by it became evident that the doubters had won the day, and in August, when he wrote to London requesting a decision on the manufacture of his smokeless powder, he met with a firm refusal.

This came as no great surprise in view of the fact that the Maxim Nordenfelt company was going through difficult times, and fortunately for Hudson he had other irons in the fire. By now he had established something of a reputation in the United States as an explosives expert, and the president of the Columbia Powder Company, impressed by his articles in the press, consulted him on the safety implications of manufacturing dynamite at their factory at Squankum, New Jersey. It did not take Hudson long to establish a productive working relationship with the Columbia company. In May 1891 he assigned to them his patent for a new 'Process of making Nitro-Cellulose', and in September he was appointed the firm's chief engineer with a salary and stock interest.

In spite of these ups and downs the brothers continued to keep one another informed about the progress of their respective smokeless powder and other experiments, although Hiram was not happy about Hudson's growing tendency to file separate American patents on lines very similar to his own. Partly for this reason, but mainly because he

---

4. *New York Times*, 21 September 1890.

wished to consult the eminent American mathematician and astronomer Samuel Pierpont Langley, he decided in the summer of 1891 to travel again to the United States, this time accompanied by Sarah. A chronic sufferer from sea-sickness, Hiram spent much of the voyage trying to devise a means of stabilising the ship's motion. The result he later patented in the form of an 'apparatus for preventing or diminishing the rolling and pitching of vessels', which was not, however, taken up.

Arriving in Maine and anxious to revisit his old haunts, Hiram took Sarah on a pilgrimage to his birthplace at Sangerville and to spend time with Harriet and Sam, with whom they were photographed in an admiring group around an 1887 model Maxim gun. They also stayed at the Haynes home in Boston, witnessing a spectacular display of kite flying from Blue Hill before going on to Washington, where Hiram was able to discuss his aeronautical researches with Professor Langley at the Smithsonian Institution. Needless to say, the couple made no attempt to see Jane and family, although they were approached by Helen Leighton, to whom Hiram passed on news of Romaine and agreed to provide more money on condition that she moved west to Colorado. Hiram appears to have met Hudson only briefly in New York before returning to England. No record has survived of their conversation, which doubtless turned on Hudson's prospective move to the Columbia Powder Company, but it is likely that Hiram made plain his displeasure about the patents situation, and that the coolness which henceforward entered into their relationship began at this time.

Nor after his return to England can Hiram have been reassured by an article on 'Smokeless Gunpowder', apparently written by Hudson, which appeared in September in the *Scientific American*. This made no reference to Hiram's contribution to the joint researches of the two men and gave the impression that Hudson alone was the inventor of the powder being developed in the United States. Hiram now openly expressed his anxiety about what was happening in America and particularly his concern to 'prevent the possibility of some outside body interfering with our business'. His fellow directors agreed, deciding that precautionary measures were called for, and in October Symon and Cloete went to New York in order to 'take the necessary steps to protect the Company's interests . . . and take charge of its business in America and Mexico'.

Hudson was now torn between loyalty to his brother and a growing awareness of the commercial possibilities opened up in the United States by the advent of the new military explosives. Nor as time went by could he see any reason why he should be inhibited from acting in pursuance of his own interests. He no longer had any contractual obligation to the company in London, and he felt free to register American patents on any inventions arising from his independent researches. After joining the Columbia company he moved from Pittsfield to New York, from where he could travel by train to the factory at Squankum, near Lakewood, New Jersey. Once again Jennie was left behind in Pittsfield to take care of the printing business, an arrangement which by now she had come increasingly to resent.

In a bid to improve his French Hudson took rooms in a boarding house run by a French landlady, who was willing to give him language instruction and to tolerate his habit of conducting experiments with explosives in a makeshift laboratory on the premises. Possibly Jennie had good reason to suspect her husband of having an ulterior motive in not bringing her to New York. Certainly she reacted strongly against this further separation, more especially after she discovered she was pregnant with their first child, and Hudson's neglect of her just when she needed his support proved fatal to their already shaky marriage. During the divorce proceedings, which seem to have gone through amicably enough, Jennie had a baby son whom she christened Hudson Day Maxim. Preoccupied with his new commitments, Hudson was not present at the birth. As a parent he was even less devoted than his brother, recording that he saw Hudson Day 'for the first time when he was a month and a half old, and again a few months later, and not afterward until twenty years had passed . . .'.

Commuting between his apartment in New York and the works at Squankum, Hudson set himself to produce a smokeless propellant which, while acceptable to the American authorities, would avoid infringing existing patents. After the Maxim Nordenfelt directors had satisfied themselves that the company's patents were secure, Hiram and Hudson patched up their differences and continued to compare notes on their work with powders based on nitrocellulose alone or in combination with nitroglycerine. Since the Anglo-German companies of the Nobel Dynamite Trust had opted for the double-based powder known to the British as cordite and the Germans as ballistite, Hudson's inclination was to go for a low-explosive powder based on nitrocellulose alone. The difference between the two categories of powder was to assume a significance of growing importance in the years to come.

After the death of Professor Mowbray in 1891 Hudson drew increasingly on the expertise of Dr Robert Schupphaus, a chemist of German origin who had formerly been Mowbray's assistant at the Xylonite company. It was already known that the fierce temperatures generated by the high nitroglycerine content of British cordite were resulting in damage to the inner linings of heavy naval guns, and so Hudson and Schupphaus looked rather to compounds of nitrocellulose more on the lines of the French BN powder which the Maxim brothers had analysed and reproduced four years earlier. Following in the footsteps of previous researchers, they concentrated on the size and shape of the grain, seeking to develop a propellant of original composition which would combine safety with ballistic performance.

Meanwhile the Maxim Nordenfelt company was making every effort to increase sales, with only limited success. Thanks to the benevolent influence of the Pax Britannica, wars were infrequent and small in scale, and the world-wide demand for automatic and quick-firing guns remained at a low ebb. In 1891 the British War Office formally adopted the .303in Maxim gun using cordite cartridges, the theory being that two of these weapons should be allotted to each battalion of the regular army. In practice, while the first official manual appeared in 1893, few guns were issued other than for training and experimental purposes. On the other hand royalty continued to take an active interest, it being reported in the summer of 1893 that the Princess of Wales had fired a Maxim gun on the Wilmington range at Dartford 'without the slightest trepidation, although the smoke was blowing in her face the whole time, which . . . took considerable courage, as the sharpness and rapidity of the explosions are distressing to most persons' ears and nerves.'

As for the services, they continued to be unresponsive. Army officers criticised the gun as complicated and expensive to operate, especially since the British soldier, using the latest magazine rifle, was capable of delivering a rapid and accurate fire with a comparatively small expenditure of ammunition. Still, too, it was seen as lacking a clearly defined role in what was termed civilised warfare. As even Capt F G Stone, a committed supporter of the Maxim gun, had to acknowledge:

> Infantrymen know their weapon and take a pride in it . . . the machine gun is nobody's child, like the Artillery of old . . . it is placed in the hands of inexperienced men, and is expected to make better shooting than the infantry soldier who is well practiced with his rifle. The tendency is not to find out what (the machine gun) can do, but rather what it cannot do.[5]

---

5. Lecture to Aldershot Military Society on 'The Maxim Machine Gun', July 1888.

After the departure of Thorsten Nordenfelt the company continued to turn out guns based on his and the Maxim patents, often of composite design. Since torpedo boats had increased in size, the Admiralty was calling for heavier weapons to deal with them, and in response a range of quick-firers was produced to compete with similar guns made by Armstrongs and Hotchkiss. By 1894, in addition to Maxim machine guns, the Erith works was turning out medium artillery in the shape of fully automatic Maxim-Nordenfelt guns up to 6-pounders and semi-automatic cannon up to 45-pounders. With an effective range of from three to seven miles, these were mainly intended for naval use, and over the years large numbers of them were bought by the French and Russian navies as well as by the British Admiralty to supplement the 3- and 6-pounder guns already mounted on ships.

For its part the British army began to find the .303in rifle calibre Maxim produced at Crayford and the Enfield Arsenal increasingly useful for policing and extending the frontiers of empire. As Lord Wolseley succeeded the Duke of Cambridge as commander-in-chief, so he was able to arrange for the weapon to demonstrate its worth in what were referred to as 'punitive expeditions' in Africa and on the north west frontier of India. In a publicity booklet issued in 1895, the company described the achievements of the Maxim gun in recent colonial wars, explaining that it gave 'a *maximum* of destructive power at a *minimum* first cost' and insisting that 'its adoption is also the most economical method of greatly increasing the effective defensive or offensive force of a country . . .'. As one of many examples it highlighted the Matabele campaign of 1893-94, when fifty troopers of the Rhodesian Chartered Company with four Maxims distinguished themselves by fighting off 5000 Matabele warriors, and a Captain Lendy of the Royal Artillery testified that 'to the Maxim guns is due, in very great measure, the success of the Company's forces. Every Matabele we spoke to told the same story. They did not mind the rifles, as they had Martinis, but what beat them off were the "Zi-go-go-gos", the name they gave to the Maxim guns.'

The weapon was indeed becoming something of a popular icon. In his memoir *Maxim Nordenfelt Days and Ways*, an employee of the company, G R Shields, related how in 1895 Sir Arthur Harris, the theatrical impresario, bought two Maxims to add spice to his melodrama 'A Life of Pleasure', at the Princess Theatre in Oxford Street:

> I remember the guns were used in a scene out East representing one of our 'little wars', and the two 'soldiers' who operated the guns were two of our mechanics . . . who were seconded for this service. The

finale of the scene was very impressive with a victory for the British of course, and the stage and the auditorium filled with the distinctive acrid smell and fumes from the black powder which was used then . . . [*ie* in blank cartridges].

But despite such publicity Maxim Nordenfelt remained heavily dependent on its overseas customers, and throughout the decade Hiram, now an internationally-known figure, found himself visiting all parts of the European continent. In company with Louis Silverman, Nicholas de Kabath or Basil Zaharoff, and occasionally Sarah, he continued to supervise trials of the weapon in Austria, Germany, Russia, Spain and Portugal, where 'the King himself fired the gun and conferred on me a high decoration'. These excursions were not without incidental diversions. In the evenings, when not accompanied by Sarah, the American was very willing to be led astray by the cosmopolitan and worldly-wise Zaharoff. The brothels of central and eastern Europe offered every variety of sexual inducement to travellers on the loose with money to spend, and it would seem that the two men freely indulged their preferences, particularly for the attentions of very young girls.

There were also complications of a different kind. In St Petersburg Hiram was interviewed by the police, who suspected from his name and that of his father that he was a Jew. On denying this, he was told that he had to register a religion, for no one was allowed to remain in Russia without one. In this delicate situation he turned to de Kabath, who suggested he call himself a Protestant. 'I asked him,' recalled Hiram, 'if a Protestant were not someone who protested against something; he admitted that such was the case. I then said to the official, "Put me down as a Protestant; I am a Protestant among Protestants; I protest against the whole thing." '

These remarks having been suitably modified in translation, the difficulty was resolved and Hiram was summoned to the Winter Palace to lunch with the Tsar and the Archduke Michael. The security police would not permit him to take a gun to the Palace, so instead he presented his hosts with 'a large and elaborately bound volume containing many photographs of Maxim guns'. Subsequently, although the Tsar himself fired the gun at the Riding School, Russian army trials produced no immediate result, but in 1892 there was a development of major importance for the future when through the agency of Lord Rothschild it was arranged for the machine tool and arms manufacturer Ludwig Loewe of Berlin to produce Maxim guns under licence. Soon it was reported that orders for '185 mitrailleuses' had been placed by the German government and in later years, thanks to the persuasive sales methods of Basil Zaharoff, the Russian army was to acquire large numbers of Maxims manufactured by Loewe.[6] Yet another effect of this agreement was to establish a family connection with the German concern when Lord Rothschild, seeking to revive the flagging fortunes of the company in which he had a substantial interest, arranged for Ludwig's younger brother Sigmund to be appointed to the board of Maxim Nordenfelt.

Hiram was now suffering increasing deafness from exposure to the discharge of firearms and artillery, and Sarah managed to persuade him to take the occasional holiday as a means of relaxing from the pressures of foreign travel, long hours in his laboratory and the stresses and strains of the factory floor. In 1890 he first learned to appreciate the sunshine and other delights offered by the south of France, and two years later he returned with his wife to pass the winter months on the Riviera. Inquisitive as always, the inventor took the opportunity to study the flight of birds and the effect of wind currents on waves, drawing conclusions which helped him in the construction of the flying machine on which he was then engaged. Visiting Basil Zaharoff at his winter residence in Monte Carlo, he was also fascinated by the Casino, where he spent hours observing the punters in action and making a careful analysis of the play.

In March 1892 Sir Gerald Graham stepped down as chairman of the company and was replaced by Admiral of the Fleet Sir Edmund Commerell VC, a change which appears not to have persuaded the Admiralty to adopt the 37mm gun. Business remained disappointing, and in the summer of the following year Sir Edmund accompanied Hiram, Zaharoff and de Kabath on a sales mission to Turkey. To Hiram's dismay they were obliged to travel by steamer from Marseilles because owing to a cholera epidemic all connections by rail were suspended. The sea, however, was calm, and on arrival in Constantinople the party presented a Maxim gun to the Sultan, who in return conferred on the inventor the Grand Order of Medjidieh and invited him to choose a lady from his harem, an offer which was, apparently, declined. He also asked them to consider, on their return journey through the Dardanelles, how best the defences of the elderly forts guarding the Straits might be improved. In due course a report was sent to the Sultan complete with proposals for strengthening the forts and arming them with quick-firing guns, but only a few orders for

---

6. Whose company was renamed in 1895 the Deutsche Waffen und Munitionsfabriken, or DWM.

Nordenfelts were placed,[7] probably because the Turkish exchequer had difficulty in paying for the Maxims and more particularly their high consumption of ammunition.

This proved a frequent obstacle to sales. The King of Denmark remarked of the 37mm gun that it would bankrupt his country in about two hours, while the Shah of Persia had to content himself with a demonstration. Hiram described how, using blank ammunition, he arranged for the rifle calibre version to be fired for the Shah's benefit in the grounds of Buckingham Palace:

> I explained to him in French how the gun worked without the aid of any manual operation, and he seemed very interested. Before I took my departure His Majesty asked me to give him the gun, but as it was worth about £220 I could not see my way to comply with this august request. At this demonstration I had some anxious moments while preventing the Shah from shooting the Grand Vizier, who would persistently lean over the muzzle of the gun whilst His Majesty was playing with the breech mechanism.

The Chinese authorities also hesitated, despite being well disposed towards the inventor on account of his attitude to the Christian missions in their country. Fung Ling of the Imperial Legation in London noted that when missionaries in Szechwan were attacked 'newspapers in all countries were full of condemnation and the English Christians were so angered that a grave situation was only narrowly averted'. In letters to the press Hiram dissented, holding that 'China, an ancient nation, had its own religious teaching and that there was no need for . . . missionaries to preach there.' When in 1896, therefore, the senior Chinese statesman Li Hung Chang came on a formal visit to England, he expressed a wish to meet the American, with whom he established an immediate rapport, and on whom he arranged to bestow his Emperor's Order of the Double Dragon.

This was not, of course, the sole reason for the visit. Since their disastrous defeat in the Sino-Japanese War two years earlier, the Chinese had been anxious to improve the efficiency of their armed services, and the purpose of Li's tour was to assess the latest weaponry being produced by the European powers. Since Maxim Nordenfelt were offering favourable trade terms, he was readily persuaded to witness the Maxim guns fired at the Eynsford range in Kent and in the grounds of the country house at Pinner in Middlesex rented by

---

7. Maxim Nordenfelt minute book, December 1893.

Sigmund Loewe, now the company's managing director. Here for the delectation of the Minister the rifle calibre gun scythed through tree trunks, while at Eynsford, before a distinguished crowd of onlookers, both versions were put through their paces. 'The bullets,' wrote Fung Ling, 'poured out like thunder rain, and afterwards the Minister was carried in his chair a great distance to see the targets made of wooden planks quite riddled . . .' But then Li inquired how much the 37mm cartridges cost, and on being told that the price was six shillings and sixpence apiece he shook his head, saying 'This gun fires altogether too fast for China'.[8]

In the United States, too, the explosives industry was going through difficult times. The scores of small companies established during the post-Civil War boom period were now in the doldrums, and mergers and takeovers were the order of the day. By the early 1890s only a few large concerns remained, and when the Columbia Powder Company tried to stand up to the might of Du Ponts they were driven out of business. With no alternative employment in prospect, Hudson decided to take the plunge and set up on his own. He had written articles and kept in touch with the Bureau of Ordnance to such good effect that his reputation as an explosives scientist was acknowledged by such luminaries as Professor P R Alger, leading specialist in naval gunnery, under whose guidance the US navy was already embarking on trials with smokeless powder. He had also sought, though without success, to interest the Department of the Navy in the wire-operated Brennan torpedo used in England for the defence of harbours,[9] and his mind was fertile with ideas for projectiles and fuses which, although mostly derived from Hiram's earlier work, he felt confident he could develop on his own initiative.

Accordingly, and helped by the association of his name with his famous brother (with whom in the United States he was frequently confused), Hudson formed his own Maxim Powder and Torpedo Company. With the collaboration of Robert Schupphaus and financial backing from the ever helpful Schuyler, he took over the Columbia company's works at Squankum, which was renamed Maxim, New Jersey. Schuyler having been appointed president, offices were opened at 41 Wall Street, and the factory continued to turn out and market a variety of explosives. At the same time Hudson and Schupphaus

---

8. Recollections of Fung Ling, Naval Attaché at the Imperial Chinese Legation, 1895-96. Vickers Archive.
9. Which its inventor had reportedly sold to the British government for the enormous sum of £100,000. In December 1886 Hiram had patented a similar device 'propelled by the winding in from the shore of a wire wound on a reel in the torpedo'.

pressed forward with their researches so that by the summer of 1894 they were well on the way to producing a multi-perforated, nitrocellulose-based smokeless powder which it was hoped would meet the rigorous specifications laid down by the American armed services. As Schupphaus prepared to patent a new process for the cheaper and more effective nitration of cellulose, the auguries seemed favourable. Then disaster struck.

The Maxim works was located sixty miles south of New York, in a sparsely populated region with a railroad connection and little else, in Hudson's words 'well situated as a place for factories that might blow up'. This was no mere flippancy. Working in his laboratory after a night disturbed by toothache, Hudson made a careless move with a test tube containing fulminate compound which exploded and blew off his left hand. Applying a rough tourniquet, he made his way with difficulty to New York, where a surgeon patched up the stump. But it was not in Hudson's nature to let a mere accident stop him in his tracks. Next day he was able to dictate his office correspondence as usual. He had an artificial hand of leather made and loosely attached at the wrist, and shortly afterwards he was jostled by a 'drunken rowdy' while descending from the elevated railway. 'I had,' wrote Hudson, 'the habit of fighting at the drop of a hat, and I struck out with my right hand and knocked him endwise. My other hand fell off and so did my hat. But I picked them up and went home greatly encouraged. I felt that I could get along in the world after all.'

After the loss of his hand Hudson had no alternative but to delegate his experimental work to others. Henceforward, as he put it, he had to depend on his wits rather than his hands, and he 'planned more and did more bossing', dedicating himself to the task of finding a buyer for the Maxim-Schupphaus powder. From the beginning of 1895 he lobbied the Navy Department in Washington and corresponded with the chief of the Bureau of Ordnance, urging the merits of his design for an automobile torpedo and of his 'cannon powder', which he offered to supply in a compound either of pure nitrocellulose or of nitrocellulose combined with a small quantity of nitroglycerine. The US government had already embarked on a battleship building programme, and both services were under considerable pressure to adopt a smokeless powder, preferably one made in America. Since the Du Pont company was not yet persuaded of the commercial viability of military explosives, it was not in a position to produce a workable military powder, even in the relatively small quantities required. For the first time the response of the authorities was positive.

In June and July 1895 Hudson was present when tests of the Maxim-Schupphaus propellant were carried out by the Bureau of Ordnance at the army's firing ranges at Sandy Hook. The results, published in August, were detailed and thorough. They were also wholly favourable, indicating that projectiles using the new powder had achieved high velocities and long range with comparatively low pressures in the gun barrel. These velocities, moreover, were obtained 'with the shorter guns of the army, and higher results are reasonably to be expected in the larger naval guns'. Summing up, the Bureau concluded that: 'The difficulties hitherto encountered in the manufacture of a smokeless powder for large guns in the form of a colloid, consisting chiefly of guncotton in the highest degree of nitration . . . seem to have been removed by Maxim and Schupphaus . . .'

Hudson, jubilant, was quick to exploit this welcome turn of events. He was by now on good terms with the editor of the *Scientific American*, to whom he sent an article describing in detail the new smokeless powder 'produced in America by Mr Hudson Maxim and Dr Robert Schupphaus, which is beginning to claim public attention from the remarkable ballistic results which have been obtained by it . . .'. This consisted of 'about 90% guncotton with 9% nitro-glycerine and 1% urea . . . a true colloid . . . made in relatively long cylinders perforated axially with a large number of small holes.' Unlike powders with a high nitroglycerine content, the Maxim-Schupphaus powder was claimed to combine performance with stability, *ie* the ability to keep without deterioration, and safety, minimising erosion to the gun barrels and the risk of premature explosions. 'At the present time nothing less than a stable explosive completely adaptable to guns of all sizes, which will produce as high ballistic results as possible . . . with a minimum of injury upon the gun, will satisfy artillerists.'

Hiram was naturally interested to hear of this signal achievement on the part of his brother, notwithstanding his reservations about what he saw as the misuse of his gun and smokeless powder patents in America. To this grievance he again drew attention when interviewed by *The Strand Magazine* in the summer of 1894:

I have not been treated well in my own country, and I am now of opinion that England offers fairer scope and opening for the inventor than the United States, where opportunities and contracts are not usually given on grounds of merit, but sold for the most advantageous terms. For this reason my gun is not adopted there, whilst my gun-carriage and smokeless powder patents have, I think I may justly say, been misappropriated . . .

This was in part a reference to the Colt Firearms Company, which, as already noted, was in process of developing the air-cooled, 'gas hammer' Browning machine gun, an initiative to which Hiram objected vigorously, protesting that it was a pirated version of his own design. Nevertheless, the Colt was a strong competitor, prompting Hiram to make improvements to the recoil mechanism of his own gun,[10] and it soon found favour with the US navy, which in 1896 was to acquire fifty, its first large purchase of automatic machine guns.

Hiram was, however, pleased that the propellant on which he and Hudson had worked together had been approved by the American authorities. At this time, and following the official adoption of cordite, Alfred Nobel was in process of suing the British government for what he alleged to be an infringement of his ballistite patent. Despite the able advocacy of John Fletcher Moulton, QC, he had failed, partly because during the trial Hiram gave evidence which was used to support the government's case: it was, declared the attorney general, Sir Richard Webster, the American and not Nobel who had been the first to combine nitroglycerine and true guncotton in a smokeless powder, and therefore the latter could claim no priority for his ballistite over the British cordite.

Despite this, Hiram had as yet been accorded no formal recognition of his pioneering work on smokeless powders, and he now wrote to congratulate his brother and to suggest that the Maxim Nordenfelt and other European companies might be interested in purchasing the Maxim-Schupphaus powder. This served to encourage Hudson in the belief that from a commercial standpoint the prospects for marketing propellant and explosive powders were better in Europe than in America, where appropriations for the armed services remained at a low level despite agitations which had led to the building of a series of battleships designed for ocean-going rather than purely coastal defence duties. In September 1895 Hudson gave instructions that the Maxim Powder Company's plant in New Jersey be extended to meet what he confidently expected to be a brisk demand, and then, armed with the Bureau of Ordnance reports on the Sandy Hook firings, he embarked on a sales mission to England, leaving Schuyler in charge of the New York office.

Back in London he found himself received rather less warmly by Hiram and Sarah, and at first he had to find his own accommodation, renting a house at 67 Effra Road, Brixton. In other respects Hiram

---

10. This was the so-called 'solid action' Maxim, in respect of which the inventor received from the gun company a special honorarium of £250.

continued to be supportive, replying to an inquiry from an English patent agency readily endorsing Hudson's claims for the smokeless powder on which he and Schupphaus had collaborated for so long:

> . . . my brother is prepared to show you the Original Official Reports from the United States. You will observe that they are better than has ever been obtained before . . . two scientific experimenters having worked at it for about six years and having spent a very large sum of money . . . The result I think you will admit is all that could be desired . . . I have no doubt that if this powder was made in large quantities our Company would be able to use a very large quantity.

In October there was more good news when a contract arrived at Effra Road from Du Ponts offering to take out an option on any smokeless propellant produced by the Maxim Powder Company. Whereas, it ran, Hudson Maxim 'claims to be the inventor and owner of certain inventions relating to smokeless powders . . . and proposes to organise an American Company for the purpose of producing his inventions . . .', he was to agree to make these available to Du Ponts in return for a retainer of $500 a month for a period of seventeen years. This he was glad to do. At last, it seemed, the years of endeavour were bearing fruit. As he opened a European office of the Maxim Powder and Torpedo Company at 55 Charing Cross Mansions in London's West End, Hudson prepared to renew his advances to the Maxim Nordenfelt directors and any other explosives company likely to be interested in buying his powder. Not that smokeless powder was his only concern. Hudson's brain was now alive with ideas which he was anxious to try out on his brother as well as the British public before bringing them to the attention of entrepreneurs in the busy European marketplace.

As usual he found Hiram engaged with several projects at once, including a new design of pneumatic rubber tyre and a process for extracting gold from refractory auriferous ores. The first trials with his flying machine (see Chapter 4) having been frustrated partly by the lack of a suitable power unit, he was trying to come up with a gasoline engine adaptable to both motor vehicles and aeroplanes. The use of gasoline or petrol was, however, severely restricted by safety concerns, while in England work on motor vehicles was inhibited by the Red Flag Law which remained in force until 1896, when the Daimler company was formed and F W Lanchester built the first British petrol-driven car. On both sides of the Atlantic the race was on to devise a safe and practicable commercial automobile.

At New Haven, Connecticut, Hiram Percy Maxim, now 26, was also working on automobiles, having already produced a powered tricycle for the Pope Manufacturing Company. While estranged from his father, Hiram Percy had remained on friendly terms with Hudson, who at one point suggested that they open a plant to produce calcium carbide, the source of acetylene, a gas which when mixed with air and ignited seemed to offer a possible alternative to benzine or petrol in the internal combustion engine. This fell through following an explosion of acetylene in the works at New Haven, but Hudson had been enthused by his nephew's belief in the future of automobiles, and he and Hiram now compared notes on the prospects for a motor vehicle for civil and military use. In November 1895 the brothers together took out a patent for 'improvements in the manufacture of pipes and tubes and in apparatus therefor', and two months later they published a joint article on 'Motors and Fuels' in the journal *The Horseless Age*. But by this time Hudson was far more excited by the outcome of a chance meeting the previous autumn which was about to turn his life upside down.

Always aware of the value of good public relations, Hudson was a frequent lecturer to learned societies and contributor to popular journals, and soon after arriving in London he was invited to address the Balloon Society on the subject of smokeless powder. His talk was followed by a lively discussion during which the pacifists present attacked the speaker as an inventor and manufacturer of war material. 'At length,' he wrote, 'a gentleman . . . distinguished looking though small in stature, rose in the audience, and in a clear, well-modulated voice took sides with me effectively. After the meeting was over I introduced myself to him, and he presented me to his wife and his daughter Lilian. He said he was a minister, a newspaper editor and author, and had written and published interviews with many distinguished men – among them Gladstone, Disraeli and Spurgeon – and asked if I would grant him an interview. I gladly acquiesced, and invited him and his companions to lunch at my apartments the following day . . .'

The gentleman turned out to be the Reverend Dr William Durban, a nonconformist clergyman who conveyed a pleasing impression of learning and refinement. Hudson, however, had eyes only for the strikingly attractive Lilian. He had already been moved to comment on the good looks and sophistication of English women, and at once he plunged deeply and permanently in love. It was a whirlwind courtship. Hudson was soon a regular visitor at the Durban home in Streatham, taking Lilian on a boat trip up the Thames to Kew

Gardens and to the Army and Navy Museum, where he was gratified by her interest in 'a wonderful display of weapons of war of every description, from the earliest forms down to the most modern types'. He even became a regular church-goer in order to see his beloved as much as possible, though he 'found the services rather irksome, and . . . watched the clock pretty well'. After three weeks he was at home reading Gray's *Elegy* with Lilian when, overcome by emotion, he reached over and embraced her, only to be told that in England a man was not supposed to kiss a girl until they were engaged. ' "My God!" I said . . . "That's just the thing for us to do. And she agreed." '

Predictably enough Lilian's parents were not altogether happy that she should commit herself quite so readily to the fast-talking American of whose background very little was known. Next day Dr Durban called on Hudson to thank him for doing them the honour of proposing to their daughter, but pointed out that it was customary in England first to ask the father's consent. This had not occurred to Hudson, who hastened to make amends, and henceforward he was accepted without question by the family. Hudson and the gentle Dr Durban became fast friends, each respecting the other's opinions despite their differences in matters of religion. Fortunately the older man was of a progressive persuasion and they found much common ground on the burning issues of the day. As Hudson wrote of his future father-in-law, there was 'no bigotry in his religion, and he and I agreed perfectly about evolution . . . he had great breadth of understanding, and his ethical ideas, ideals and philosophy of life were scientifically righteous.'

As for Lilian, she was deterred neither by Hudson's artificial hand nor by the knowledge that he had a son by a previous marriage. She appears to have fallen without reservation for the ebullient American, who, with his rakish charm and quirky sense of humour, came into her life like a breath of fresh air and swept her off her feet. From the beginning each recognised that given their different viewpoints concessions would have to be made, but more usually it was Lilian who gave way gracefully to the forceful Hudson. It was agreed, he wrote, 'that she would go to the theatre with me as often as I went to church with her', while he thoughtfully arranged for her to read to him in the evenings Haeckel's *History of Creation*, 'a book that makes unquestionably clear the fact of evolution. After about two weeks she exclaimed, as she was reading, "Why, Hudson, this is true!" "Of course it's true," I said.'

In March 1896 Hudson and Lilian were united in 'a regular English church wedding', spending their honeymoon in Brighton. To an

extent the occasion helped to smooth over the awkwardness between
the brothers and afterwards, following the Maxim family tradition, the
couple went to live at Thurlow Lodge, to which Hiram and Sarah had
returned following the conclusion of the inventor's flying machine
experiments in Kent. Sarah and Lilian were soon on the best of terms,
and Hudson was glad of the opportunity to help his brother complete
and equip a laboratory where they intended to work on the various
projects that engaged them both. At the same time the younger man
sent out letters (headed in characteristic style 'hMuAdXsIoMn') to
individuals and commercial companies he thought might be interested
in the smokeless powder and other inventions which his company was
concerned to exploit.

In May Hudson summed up his contribution to the development of
high explosives and smokeless powder in an article in the Journal of
the Society of Arts. Since beginning his investigations in England
eight years earlier, he wrote, 'I have built two powder mills in America, one for the manufacture of high explosive, the other for smokeless
powders for ordnance . . . I conducted a large number of experiments
for the . . . Zalinski Dynamite Gun Co of New York . . . for the
purpose of throwing high explosives or aerial torpedoes from ordnance.' More recently he had organised the Maxim Powder and Torpedo Company 'to operate under my inventions and to carry on with
this work', despite the misguided criticism of pacifists and others who
objected to the manufacture of war material. 'At best,' he asserted,
making a point to which both brothers were constantly to return, 'war
is cruelty – yet it is often a necessity, and once engaged in should be
made as terrible and destructive as possible . . . in order that it may be
as brief as possible, thus minimising the evil . . . The most deadly and
destructive engines of war are the most humane, and those employed
in the production and development of them may justly be looked upon
as humanitarians.'

Three months later Hudson wrote somewhat disingenuously to his
mother back home in Wayne to inform her and Sam about his marriage to Lilian. Harriet had kept in touch with Jennie and baby Hudson and was no doubt pleased to have his news, however belated : 'Her
father is a clergyman, but his principal business is writing for periodi-
cals . . . He is a great linguist and is now travelling in Russia.' He and
his new wife were staying at Thurlow Lodge, 'the old place where Hi
used to live . . . We are building a very fine laboratory here, where
Hi and I are going to conduct a lot of experiments. How old is little
Hudson now? And does he take any interest in scientific things? I have
not seen my little boy for about two years, but I saw Jennie the last

time I was in New York . . . I hear from her often, however, and gave her some money for the boy . . .'

By the summer of 1896 Hudson was no nearer to finding potential customers for his smokeless powder and other inventions. Du Ponts sat on the fence, waiting for the United States authorities to place firm orders for the new explosives, and in New York the shareholders of the Maxim Powder and Torpedo Company were becoming anxious. Hudson still hoped that Maxim Nordenfelt might be willing to adopt the Maxim-Schupphaus powder, but he was not to know that behind the scenes changes were taking place which made this well-nigh impossible. A few months earlier Sigmund Loewe had taken overall charge of the company's affairs, and at once he embarked on a programme of reorganisation and retrenchment. Described by the historian of Vickers as 'a small, fierce, bearded, vibrantly energetic man, who alarmed and magnetised his staff, shaking them by his rages and charming them by the discrimination of his praise',[11] Loewe wielded the new broom with vigour and was no respecter of persons or of reputations. One of his first steps on becoming director had been to propose that the services and financial guarantees enjoyed by Basil Zaharoff be temporarily suspended, and soon he was questioning every aspect of the company's operations. Whether directly or indirectly, the reforms which he proceeded to drive through were fundamentally to affect the careers of both the Maxim brothers.

Loewe shared the reservations of several of his fellow directors about many of Hiram's activities. No one doubted that the American's standing as an inventor made him indispensable, and indeed in June 1895 he was reappointed as 'Director and Engineer' for a further period of seven years. There was, however, no denying that he could be awkward to deal with and was, at £2000 a year, an expensive asset. In the volume of reminiscences already quoted, G R Shields described Hiram at this time as 'rather distinguished looking with his snow white hair and beard. He was exceedingly deaf, and it was amusing to see Mr Cloete holding the lobe of his ear and speaking into it as if using a speaking tube. Maxim's deafness did not stop him talking; he would hold forth on the slightest provocation on subjects from pumice stone (on which he gave me a long lecture in the clerk's lavatory one day) to the latest ideas in aeronautics.' His 'inventor's itch' also tended to have an adverse effect on the daily routine of the factory, so much so that the directors 'barred him from entering the Erith Works and

---

11. J D Scott, *Vickers, a History* (London 1962), p45.

allocated a shop and sufficient machines at Crayford where he could work away at his ideas without interfering with anyone else.'

Shields noticed that besides having 'a peculiar sense of humour', Hiram was apt to irritate his more conventional colleagues by playing to the public gallery, a good example of this being the strange affair of the 'Doewe cuirass'. In May 1894 a Berlin tailor, Heinrich Doewe, hired the Alhambra Theatre in London, where he and an accomplice, one Leon Martin, supposedly a US army officer, demonstrated a bulletproof jacket apparently fabricated of cloth, canvas and buckram to gatherings of interested spectators, among them HRH the Duke of Cambridge and his staff. In essence their act consisted of Herr Doewe standing on stage wearing the jacket, or cuirass, and Captain Martin firing several shots at him from a Lee-Metford service rifle. As Herr Doewe walked off the stage, so the deformed bullets dropped out of the jacket and on to the floor. A well known rifle shot, Frederick Lowe, examined the bullets and, finding them still hot, was convinced that the demonstration was genuine.

Hiram was equally convinced that it was a fraud. As it happened, he had been working on steel shields to protect the men operating his machine gun against rifle fire, and he concluded that Herr Doewe had deceived his audiences by wearing a piece of chrome steel plate concealed in padding underneath the cuirass. Accordingly, having prepared a 'jacket' concealing a plate of nickel steel, he wrote to the newspapers dismissing the performances at the Alhambra as a 'howling farce'. He also announced that he had devised a bulletproof garment which was lighter than Herr Doewe's but equally effective, and he invited journalists to attend a demonstration at Erith. Hiram duly astonished what he described as an 'immense crowd' of pressmen by showing that he could be fired at with impunity without any of the bullets passing through his clothing. He then discomfited his audience by revealing how the trick was done. After that, he commented, 'Herr Doewe's bullet proof cuirass did not meet with any further success . . . [though] in the meantime he and his confederate had made a lot of money out of it.'

While no doubt affording Hiram much innocent satisfaction, incidents such as this did little to reassure Sigmund Loewe that the American was pulling his weight in the struggle to put the company's finances on a sound footing. In particular Loewe was critical of the expense involved in Hiram's flying machine experiments, which, although promising much, seemed unlikely to yield any commercial return at least in the short term. He therefore urged Hiram to look for means of capitalising on those of his patents that were already established, and in particular those connected with ordnance and explosives. One suggestion was that the company seek to profit from Hiram's years of research into smokeless powder by initiating legal proceedings against the British government on the basis that his patents of 1888 had preceded that of Dewar and Abel's cordite. This the government had acknowledged during the great cordite trial of 1894-95, when it was used in argument to resist the action brought by Alfred Nobel. Now the hope was that Hiram might in his turn challenge the monopoly position enjoyed by cordite, so opening the way for the manufacture by Maxim Nordenfelt of the inventor's Maximite as an alternative propellant for military use.

By the summer of 1896 the company's lawyers were engaged in preparing its case against the government, and this being so it is not surprising that Hudson met with little positive response to his overtures to Sigmund Loewe in the matter of the Maxim-Schupphaus powder. Loewe was pleased to see Hiram's resourceful brother, of whom he had heard so much, and to accept samples of his powder. He listened politely to Hudson's proposal that Maxim Nordenfelt should pay for trials with a view to its manufacture in England. But a week or two later he wrote pointing out that the British authorities were committed to cordite and that there was as yet no indication of their being willing to consider any other powder. He denied that his firm had entered into any agreement with Hudson, adding bluntly that: '. . . we have no interest in getting your powder introduced into this country, and we certainly cannot do so at our expense.'

Angry and frustrated, Hudson replied citing his brother's support and insisting that Maxim Nordenfelt had indicated their wish to place 'a regular order for the powder, agreeing to pay five shillings a pound for it'; consequently a consignment had been put together and was in America awaiting shipment. At Hiram's suggestion he also wrote to the Vickers company, which offered to collaborate with Maxim Nordenfelt in carrying out tests on his smokeless powder at their artillery range at Swanley in Kent. But further delays followed, so that by the autumn Hudson had given up all hope of reaching an accommodation with either Vickers or Maxim Nordenfelt. Instead he explored alternative possibilities, making inquiries with a view to setting up his own manufacturing plant in England and sending information about his powder to other explosives concerns including the Hotchkiss company in Paris and Armstrong's Elswick factory at Newcastle. The former declined Hudson's offer because, they said, the government 'monopolises the fabrication of all explosives in France and we are hardly prepared to establish other Continental works'. More encouragingly,

Sir Andrew Noble of Armstrongs indicated his firm's interest and asked for a meeting in London.

Hudson also pressed ahead with other projects he had discussed earlier with Hiram and Hiram Percy Maxim. In November he formed the Maxim Carbide and Acetylene Gas Syndicate of Queen Victoria Street, London, which it was hoped 'will virtually control the carbide business'. Hiram was already a founder member of the Motor Car Club (later the Royal Automobile Club), established 'to promote the manufacture and use of motor carriages and other road vehicles', and a month or two later the brothers together opened negotiations to float the Maxim Auto-Car Syndicate. Hudson wrote to his nephew in America inviting him to join the venture, but, essentially good natured as he was, Hiram Percy saw no reason to overlook his father's callous treatment of his mother and sisters and he declined to become involved: 'I have,' he replied, 'long ago decided . . . that Hiram is unworthy the attention of either a man conversant with logic or one who is willing to stand upon his own resources, and that the less one has to do with him the better . . .'

Thanks to the dubious but effective sales methods of Basil Zaharoff and the administrative changes brought about by Sigmund Loewe, the fortunes of the Maxim Nordenfelt company began steadily to improve. Having lost £21,000 in 1894, the company made a profit of £64,000 in 1895 and £138,000 in 1896. In any case Lord Rothschild had now decided that the time was ripe to negotiate a series of mergers and take-overs designed to unite the Vickers family's substantial interests in steel, shipping and armaments. At a time of growing international tension Great Britain was diplomatically isolated and public opinion, spearheaded by the Navy League, was calling for action to strengthen the Royal Navy. As the naval estimates for 1896-7 swelled to nearly £22 million, Vickers' purchase of Maxim Nordenfelt and the Naval Construction and Armaments Company at Barrow[12] was nothing if not timely. So was created the great industrial conglomerate of Vickers, Sons and Maxim which was to become, together with Armstrong-Whitworth, the leading British supplier of munitions to the armies and navies of foreign powers and the bulwark of British naval might during the period up to and during the First World War.

---

12. Where two of Thorsten Nordenfelt's submarines had been built and warships were already being equipped with Nordenfelt quick-firing guns.

FOUR

Gas-bags and Flying Machines

It has certainly not been made easy for human beings to travel the realms of the air freely like a bird . . . Unfortunately progress is hardly ever made beyond the first attempt which normally ends either by failure to rise in the air, or, if this is achieved, by inability to land with undamaged apparatus.

OTTO LILIENTHAL (1895)

No single manifestation of Hiram Maxim's vision and resource as an inventor is so remarkable, and none has aroused so much argument, as his contribution to the history of manned flight. All the evidence indicates that most of his contemporaries and fellow workers in the field saw him as an important innovator, marking out a trail which others were soon to follow. More recently, and with the benefit of hindsight, the value of his pioneering work has been called in question and even dismissed out of hand, but this is a matter of controversy on which expert opinion is still, and perhaps always will be, divided.

Since time immemorial men had been trying to take to the air. The Chinese were the first to use kites, the archetypal aeroplane, as an aerial vehicle, while Europeans were the first to develop the windmill, the origin of the propulsive airscrew. By 1850 experimenters in Europe and America were grappling with the problems involved in achieving controlled flight by means either of an airship or of a heavier-than-air machine. The contest between the advocates of the two systems was close and was to remain so until well into the twentieth century. The lighter-than-air enthusiasts opened up an early lead, for hot-air and gas balloons had been familiar since they were introduced by the Montgolfier brothers at the end of the eighteenth century. The first successful non-rigid airship, designed by the Frenchman Henri Giffard, caused a sensation when it flew over Paris in 1852, and observation balloons were used extensively during the American Civil and Franco-Prussian Wars. Heavier-than-air machines, whether in the shape of the ornithopter, driven by bird-like flapping wings, the helicopter or the aeroplane, were bound to be a more difficult proposition.

Hiram summed up his views on aviation in the most impressive of his printed works, *Artificial and Natural Flight*, published in 1908. This reveals a fascination with the subject which began, like his interest in the machine gun, fifty years earlier when he was out walking with his father in the woods of Maine. From his observation of birds and other natural phenomena, Isaac concluded that the future lay not with balloons but with the heavier-than-air machine. Hiram was equally convinced, holding consistently to the belief that: 'In all Nature, we do not find a single balloon. All Nature's flying machines are heavier than the air, and depend altogether upon the development of dynamic energy.' Isaac's own preference was for a helicopter on the lines of the string-pull toy popular with generations of children, but of course at that time no source of energy existed which was sufficiently light yet powerful enough to drive anything larger into the air.

Ever since, Hiram had at the back of his mind the possibility that one day he might make a flying machine that could lift itself and its operator off the ground under its own power. It was an ambition he shared with many others. In his introduction to Mottelay's biography Lord Moulton states that the American 'rarely, if ever, took the trouble to ascertain the existing state of knowledge in any subject which he took in hand . . . [and so] he assumed that he was the first and true inventor of it.' But this is contradicted by Hiram's repeated assertion that it was his practice to read extensively around any subject that engaged his attention, and that in the context of his flying experiments he 'procured all the literature available . . . both English and French, and attempted to make a thorough study of the question' before going ahead. His own technical writings[1] make it clear that he was familiar with the work of most if not all of his many precursors, as he was aware of the contemporary flying experiments being carried on by Clément Ader in France and Samuel Pierpont Langley in the United States.

Not until 1887, when he was already busy with other projects, did Hiram finally take up the challenge of flight. Like many of the early pioneers he did so in the expectation of military contracts, and from the beginning his aim was to produce a machine with military capability rather than for transporting passengers or freight. The initial stimulus came from Symon and Cloete, his colleagues in the gun company. The Maxim gun having been effectively launched, they were looking for other means of capitalising on Hiram's inventive skills, and the notion of producing an aircraft for use in war was one which held out

---

1. Notably his paper *Progress in Aerial Navigation* (c1890) and a detailed *Report on Aerial Machines, 1842-1893* in the Maxim Collection, Connecticut State Library.

'...a thing that must be done'. Edward, Prince of Wales, tries his hand at the Maxim gun, watched by Hiram, Louis Silverman and a group of interested spectators. Note the cloud of black powder smoke. (Collection Dolf Goldsmith)

Rifle calibre Maxim for field use. (Vickers Archive)

(No Model.)

H. MAXIM.
STEAM COOKER.

No. 293,048.

Patented Feb. 5, 1884.

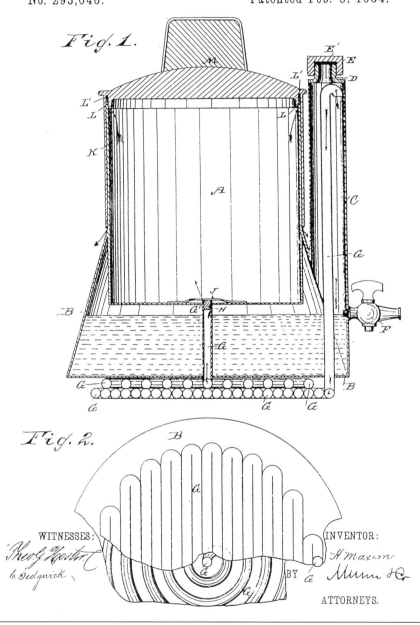

Fig. 1.

Fig. 2.

WITNESSES:

INVENTOR:

H Maxim

BY

Munn & Co

ATTORNEYS.

The first indication that Hudson, too, was a budding inventor. This early attempt at a pressure cooker was not, however, a success. (Hudson Maxim Papers)

Above: Hiram, already white-haired from contact with the toxic chemicals used in explosives, instructs Hudson in the working of the Maxim gun, Bexley Heath, 1888. The only known photograph of the brothers together. (Hagley Museum and Library)

Left: Demonstrating the power of the Maxim gun, Pinner, 1896. Hiram stands at right with His Excellency Li Hung Chang. Albert Vickers and Sigmund Loewe at left with Fung Ling of the Chinese Legation. (Vickers Archive)

Hiram and his second wife Sarah (on right) with Harriet, Hiram's mother, and his brother Sam during their visit to Maine in 1891. The Maxim gun is an 1887 world standard model, one of the first to be shipped to the United States for service trials. (Collection of Val J Forgett Jr)

Left: Hudson's first wife, Jennie Morrow Maxim, with their son Hudson Day Maxim, 1891. (Hudson Maxim Papers) *Right*: Lilian Durban at about the time of her first meeting with Hudson. This portrait always took pride of place in the inventor's library. (Lake Hopatcong Historical Society)

great attractions. In *My Life* Hiram recounts how his fellow directors asked him 'if it would be possible to make a flying-machine that would fly by dynamic energy without a gas-bag, how long it would take, and how much it would cost.' Never one to underestimate his own powers, the American was confident, giving it as his opinion that if a domestic goose could fly then so could a man, and he declared that he could make such a machine given five years and a budget of £50,000. The first three years, he said, would be devoted to the development of a light and quick-running internal combustion engine, the rest to making experiments and building the machine.

This estimate being far beyond the means of the gun company, the idea was shelved until it was raised again by Symon and Cloete after the formation of Maxim Nordenfelt in July 1888. To the directors of the new concern it seemed a far from practicable proposition. Financially they were fully committed to the business of managing the gun factories at Erith and Crayford, and the majority view was that Hiram should confine himself to more urgent priorities. Others, including Albert Vickers, were so impressed by their colleague's assured manner and his apparent ability to solve the most intractable of technical problems that they were inclined to give him his head. All, however doubtful, were intrigued by the potential for military purposes of a flying machine and the commercial prospects that such a machine would open up. The result was a compromise whereby Albert Vickers, Symon, Cloete and other of Hiram's admirers agreed to put up the sum of £20,000 over a five-year period to enable the inventor to pursue his researches, albeit on a strictly part-time basis, into what he referred to as 'the first Kite of War'.

Hiram was already a member of the Aeronautical Society of Great Britain, the first such body to be established anywhere, and now he devoted what little time could be spared from his other activities to preliminary study. As it happens he was in the right place, for whereas the French had taken the lead with balloons and airships, the British had long been noted, and frequently derided, for their preparedness to embark on more or less eccentric projects involving heavier-than-air machines. Throughout 1888 he investigated the successes and failures of these early researchers. The works of Sir George Cayley, republished in 1876, outlined the basic principles of aerodynamics and described experiments with man-carrying kites and lifting surfaces flown on the ends of 'whirling arms'. Hiram learned of the pioneering efforts of W S Henson and John Stringfellow, of miniature monoplanes launched from wires, their 'aerial steam carriage' and a much publicised but wholly impractical scheme for an 'Aerial Transit Company'. From the

marine engineer F H Wenham, who in 1866 had delivered the address
at the inaugural meeting of the Aeronautical Society,[2] and Horatio
Phillips, who specialised in the study of aerofoils, he acquired much
useful information on superposed wings and the use of wind tunnels to
measure airflow and the forces of lift and resistance.

Unlike many researchers who were preoccupied with mathematical
calculations and formulae, Hiram decided from the outset to discard
the results of purely theoretical work on the grounds that no two
mathematicians were able to agree on any one point. As he put it, 'I
think we might put down all of their results, add them together, and
then divide by the number of mathematicians, and thus find the average coefficient of error.' Instead, always the pragmatist, he started
from first principles, assuming that success could come only from
conducting his own experiments, making his own observations and
drawing his own conclusions. In view of the shortage of time and
money the inventor set himself a precise and limited objective. 'When
I made my experiments,' he wrote, 'I only had in mind the obtaining of
correct data, to enable me to build a [manned] flying machine that
would lift itself from the ground. At that time I was extremely busy,
and during the first two years of my experimental work, I was out of
England fourteen months.'

It was a tall order. Hitherto no heavier-than-air craft other than a
model had ever succeeded in lifting itself from the ground and flying
under its own power, and even such basic principles as the working of
the screw propeller were barely understood. As Edward Hewitt, who
was engaged for a while on the project, was to note, '. . . there was
almost nothing known . . . about the action of screw propellers in air,
and it was even maintained by many engineers that little power could
be developed by a screw in an elastic medium like air, [so] it was
necessary to get at all the facts accurately from the beginning . . .'[3]

At an early stage Hiram rejected the ornithopter with its flapping
mechanism and the helicopter with its horizontal rotating blades in
favour of a machine with superposed wings or aerofoils (or, as he
called them, 'aeroplanes') driven by screw propellers. Obviously the
whole framework would have to be made of the lightest possible material, and Hiram was encouraged to find on one of his visits to France
a firm making thin steel tubing which seemed ideal for the purpose.
More problematical was how to devise a lightweight engine capable of

2. On 'Aerial locomotion and the laws by which heavy bodies impelled through the air are
   sustained'.
3. E R Hewitt, op cit, p122.

generating enough thrust to lift the machine into the air. Hiram's first thought was to use internal combustion engines of the Brayton or Otto type, but these were still unreliable and relatively low-powered and, since he had no time to work on improving them, he decided to settle for the well-established and to him more familiar technology of steam.

During the course of 1889 Hiram applied the engineering expertise he had acquired over the years to the design of a steam engine using a boiler heated by naphtha gas, an ingenious arrangement of copper tubes which *The Engineer* likened to the Thornycroft or Yarrow marine boilers. Having filed his first patent for 'a lightweight steam aero-engine', this was soon followed by another relating to a 'basic flying machine', an all-metal structure with a kite-like centre section and driven by twin steam engines, and the Maxim Nordenfelt board gave him the green light to go ahead with its construction. Accordingly he rented Baldwyn's Park, a mansion set in forty acres of rolling grassland at Bexley Heath, recruited two American mechanics and set about erecting a large timber building in which to house the machine. This, the first 'hangar' to be built anywhere, the inventor thought necessary both for protection against the weather and to shield his activities from the prying eyes of an inquisitive public. While the company's directors were inclined to be security-conscious, Hiram was all too aware that experimenters with flying machines were liable to be bracketed in the same category as cranks trying to discover the secret of perpetual motion or the Philosopher's Stone.

Throughout 1890 the inventor planned and supervised a series of experiments to determine the nature and design of his machine. From an early stage he corresponded with S P Langley, secretary of the Smithsonian Institution in Washington, who, with the help of official funding and a full-time staff, had been working with small rubber-driven monoplanes and was now planning a series of more ambitious steam-powered models, or 'Aerodromes', launched by catapult. Accordingly Hiram devised a larger version of Langley's 'whirling table', which he described as 'a species of merry-go-round, that is a very long and slender arm made to travel round a large circle'. Also driven by steam, this was used to observe and measure the relative efficiency of different curvatures of wings, the porosities of fabrics and the pitch and shape of the airscrews needed to propel the machine forwards and upwards. Careful attention was paid to the most suitable material for making the propellers and for covering the wing surfaces.

Hiram soon reached the conclusion that because of the size and weight of the steam engines needed to provide the requisite power, the whole structure of his machine would have to be on a much bigger scale than anything previously contemplated. Already he was encountering a good deal of scepticism, his application for an American patent being turned down on the ground 'that the machine was not provided with a gas-bag, and consequently had no means of lifting itself . . .'. Nevertheless, on his visit to the United States in November 1890 he wrote to the *New York Times* declaring that 'within a very few years some one . . . will have made a machine that can be guided through the air, will travel with considerable velocity, and will be sufficiently under control to be used for military purposes . . .'. 'I am satisfied,' he added, ' that when the problem is solved it will be an American who solves it . . .'.

The inventor was also sufficiently confident of success to cultivate his earlier acquaintance with George Leveson-Gower, Duke of Sutherland, himself a founder member of the Aeronautical Society, whom he invited to Baldwyn's Park to inspect progress. Having provided himself 'with very costly and accurate apparatus,' he wrote, 'I have gone into the subject [of aerial navigation] in a more thorough manner than any of my predecessors, and where other men have speculated and written books in regard to what they thought to be true, I have actually tried the experiments, and I now know exactly the force required for flying . . .'[4]

The Duke, who according to the *Dictionary of National Biography* was 'fond of riding on locomotive engines', had spent large sums on the building of the Highland railway, and he had also offered a prize of £100 'to any [aviator] who would make a flight to the top of Stafford House', his imposing mansion overlooking London's Green Park. Now he was able to see for himself how far Hiram had progressed, and he may have helped with a practical difficulty exercising the inventor's mind, namely how to manoeuvre his unwieldy machine on an estate dotted with large trees. On renting Baldwyn's Park he had obtained a verbal assurance from the landlords that he could cut down any of the trees on the estate that happened to impede his experiments. When he asked permission to do so, however, a fee of £200 per tree was demanded, which was out of the question. The alternative solution adopted by Hiram was to build a railway track between the trees on which the machine could be moved in and out of its hangar, and which could be used both for controlled trial runs and the eventual attempts at 'flight'.

Over the next three years Hiram and a dedicated team of mechanics and craftsmen applied their skills to the flying machine and the track

---

4. Letter of 3 December 1890, Duke of Sutherland's papers, Stafford County Record Office.

on which it was to operate. Predictably enough, the technology on which they worked derived largely from marine and railway engineering, which was not always a reliable guide. Thus the American's experience of shipbuilding in New York led him to assume that the propellers be located at the back, to push the machine forward, and he continued to insist that: 'What is true of ships is true of flying machines . . . Good results can never be obtained by placing the screw in front instead of in the rear of the machine', since the backwash would have a 'decidedly retarding action'.[5] By a similar process of reasoning, Hiram regarded the sheer power of the engines as all-important: 'Without doubt the motor is the chief thing to be considered. Scientists have long said, Give us a motor and we will very soon give you a successful flying machine.'

Needless to say all was not plain sailing, and there were constant mishaps and breakdowns. As with the machine gun, the inventor produced a first model which was found unsatisfactory, followed by a series of improved versions, the last of which he believed to be the best possible given the nature of the compound steam engines employed. As finally developed, each of these generated 180 brake horsepower and weighed only 320lb; in the words of one of Hiram's later co-workers, the young engineer Albert Thurston, they were 'the best that had ever been done, and appeared marvellous at the time'.[6]

Hiram's initiative soon aroused interest abroad. In February 1891 the French-born American civil engineer Octave Chanute, whose hobby it was to record and encourage attempts at aerial navigation worldwide, wrote from Chicago asking for details, so initiating a lengthy correspondence between the two men. Later that year Hiram went to Washington to compare notes with Professor Langley, and on his return he concentrated on the building of a 9-foot gauge railway line 1800 feet in length, the idea being that as the machine ran down the track and 'took off' the points at which it did so could be measured by means of red paint on the wheels. After the first runs had shown that at speed it was impossible to keep the machine on the track, additional safety rails of Georgia pine were erected 13 feet outside and above the level of the original rails. The machine was then equipped with springs so that any lift could be measured, and four wheels,

5. *Artificial and Natural Flight*, p49.
6. A P Thurston, *Elementary Aeronautics, or the Science and Practice of Aerial Machines* (London and New York 1911). One engine is on display in the Science Museum, Kensington. The other is said, though without firm evidence, to have gone down in the *Titanic* while being delivered to a museum in the United States.

mounted on outriggers, were fitted in such a way that on lifting they engaged with the outer, retaining track.

As Hiram directed more time and energy to his flying experiments, those directors of the Maxim Nordenfelt company who regarded the exercise as an unnecessary distraction began to grow restless. In July 1891, with Hiram in the United States, questions were raised at a board meeting 'as to the duties of Mr H S Maxim', and in November the inventor was asked to resign as managing director of the Crayford works, being appointed instead 'Engineer' at the same salary, his duties being confined to 'making inventions'. Due to legal complications this change in Hiram's status was quickly rescinded, but over the following months there were more complaints about the rising cost of labour and fuel at Baldwyn's Park. The response of the American was deliberately to draw attention to his experiments by writing to the newspapers and inviting small parties of influential people to Baldwyn's Park to see for themselves what was being done.

It was not long before the press was sitting up and taking notice. Some of the comment that appeared was ill-informed, some frankly ribald in tone, and Hiram decided it was better to court publicity than shun it. In October an article in *Century Illustrated* on 'Aerial Navigation, the Power Required' attracted much attention, as did a further piece in *Cosmopolitan* magazine the following June, which, entitled 'The Aerodrome', described in detail the nature and the purpose of his flying machine. 'Professor Langley and myself,' wrote Hiram, 'are both independently engaged . . . on similar experiments, with a view to finding out how much power is required for flight.' He also outlined the procedure to be adopted when testing his machine's kite-like central section:

> We should first begin by running slowly – say at a rate of twenty miles per hour – and carefully note the lift on the indexes over each wheel. If we found that . . . three-fourths of the load was lifted off the forward axletree and only one fourth off the hind one, then we should change the centre of weight farther forward . . . We should then make another trial and if we found that the lift was equal both fore and aft we should increase the speed very carefully . . . until the whole weight of the machine was supported by the aeroplane, and the whole weight of the wheels – about one ton – by the railway track. Then when there was neither lift nor load on either wheel, we might . . . turn our attention to the subject of steering.

Hiram went on to describe what he expected would happen after the machine lifted into the air, although not surprisingly this was treated only in the most general terms. Just the same it did appear that his experiments were more likely to result in some practical outcome than any others then going forward. Visiting Baldwyn's Park during a European tour in the summer of 1892, Langley was impressed by the scale and thoroughness of Hiram's preparations: 'I saw Maxim and his machine,' he reported to Chanute. 'The whole thing looks very near complete. This is a serious affair . . . not to be named in the same connection as Ader and Trouvé [in France]'.[7]

About this time Hiram and Albert Vickers and other colleagues who continued to have faith in the enterprise drew encouragement from news filtering across the Channel. In October 1890 the electrical engineer Clément Ader (who later gave the word 'avion' to the French language) had taken off in his steam-powered, bat-shaped monoplane, the *Eole*, or 'god of the winds', and reportedly travelled some 8 inches above the ground for a distance of 160 feet. Though working on different lines, Ader appeared to be one step ahead of Hiram. The *Eole* was claimed as the first piloted machine to raise itself from the ground, and in 1892, on the strength of its success, the inventor received a subsidy from the French War Ministry to build a new and improved aircraft suitable for military purposes, that is, for scouting and reconnaissance. To Maxim Nordenfelt it was welcome confirmation that if a serviceable flying machine could be produced, government contracts would not be far to seek.

With the assistance of his chief mechanics Thomas Jackson and Arthur Guthrie, Hiram now embarked on a sequence of test runs along the track. He liked to take the controls himself and was normally accompanied by an engineer and at least one passenger. No attempt was made to lift off from the rails, for the first objective was to assess the performance of the engines and propellers and the effect of elevators and 'aeroplanes'. But the machine was clumsy and accident-prone, and in October 1892 the inventor, discouraged by a series of expensive setbacks and irritated by the negative attitude of some of his colleagues, wrote to Chanute asking whether it might be advisable for him to exhibit his apparatus at Chicago during the 1893 World's Fair, 'the cost to be covered by throwing open the machine and workshop to the American press and public'. Hiram's further thought was that his experiments might be continued in the United States, but while this idea appealed to Chanute he had reservations about the commercial feasibility of exhibiting a flying machine 'which has not yet flown', and the proposal came to nothing.

---

7. Letter of 2 August 1892, Octave Chanute Papers, Library of Congress.

Another early visitor to Baldwyn's Park was the budding balloon enthusiast Griffith Brewer, later to become a close associate of the Wright brothers and their patent agent in the United Kingdom. As he recalled long afterwards: 'One of my earliest aeronautical experiences was that of landing from a balloon near the Park and being entertained to tea and listening to the explanation of the machine. If anyone of that time . . . was as an engineer qualified to build such a machine it surely was that inventive genius H S Maxim . . . I returned to London almost persuaded of Maxim's success, but not sufficiently convinced to take any steps in anticipation of passing from our sport of ballooning to the days of flight by aeroplane.'[8] 'There can be no doubt,' he wrote in a piece for the *Morning Post*, 'that Mr Maxim's machine is the nearest approach to a practical flying machine yet made, and even if he does not succeed in flying he will have advanced the science of aeronautics to a considerable extent.' This was the signal for a number of journalists to travel down to Kent, and various articles, many of them written or inspired by Hiram himself, appeared in the newspapers as well as in magazines ranging from the *Model Engineer and Electrician* to *The Engineer* and the *Journal of the Society of Arts*. When the aeroplane was perfected, the inventor was reported as saying, 'the whole world will be changed . . . there will be no more ironclads, no more guns, no more fortifications, no more armies. There will be no way of guarding against what this machine will do'.

This was heady stuff, and indeed those taken for a run were not likely to forget the experience. Bucknall Smith of *The Strand Magazine* wrote of 'the indescribable sensation of mixed exhilaration and trepidation, on rushing off at a speed of fully forty miles an hour on the bosom of a veritable hurricane of this mechanical bird's own manufacture.' An American journalist, H J W Dam, ('The Maxim Air-Ship: an Interview with the Inventor') left a more detailed account:

> I mounted the platform, made of light matchboard so thin that it seemed scarcely able to bear a man's weight. Prior to the start a rope running to a fixed dynamometer was attached behind to measure the forward impulse or push of the screws. The action of these screws caused very little shaking . . . and this was a surprise to me considering the tremendous force within so delicate a framework. Ten feet behind the ship two men were shouting from the dynamometer, and indicating the push on a large board for the engineer to read. The index quickly marked in succession 400, 500, 600, 700 and finally

8. Griffith Brewer, *50 Years of Flying* (London 1946), p86.

1200lb of push and then the Commander [Hiram himself] yelled 'Let go!' A rope was pulled, and the machine shot forward like a railway locomotive – and with the big wheels [*ie* propellers] whirling, the steam hissing, and the waste pipe puffing and gurgling, flew over the 1800 feet of track. It was stopped by a couple of ropes stretched across the track working on capstans fitted with reverse fans. The stoppage was quite gentle. The ship was then pushed over the track by the men, it not being built, any more than a bird, to fly backwards.

Once again, as in the case of the machine gun, Hiram entertained a succession of distinguished visitors anxious to witness a demonstration of the remarkable new machine and be given a ride on the footplate or 'deck'. These included the eminent scientist Lord Rayleigh, the barrister J F Moulton,[9] the Chinese minister Kung Chao-yuan, the writer H G Wells, Basil Zaharoff and the new chairman of Maxim Nordenfelt, Admiral Sir Edmund Commerell. All came away impressed by the inventor's determination and by the vigour of his response to the challenge of flying, even if they were not all as confident of his chances of success.

As time went by so the doubts began to multiply. During the summer of 1893, while making the journey by steamer from Marseilles to Constantinople and back, Hiram was struck by the effortless, soaring flight of seagulls. The bird, he concluded perceptively, 'seeks out an ascending column of air, and while sustaining itself at the same height . . . without any muscular exertion, it is in reality falling at a considerable velocity though the air that surrounds it.' His machine was a long way from emulating this example. In February he had replied to an inquiry from Chanute by describing yet another accident with his apparatus when it was tipped off the rails by a sudden gust of wind and the propellers damaged the central section. This involved a further delay and deepened the reservations felt by many of Hiram's colleagues as well as members of the Aeronautical Society, whose projected visit to Baldwyn's Park had again to be postponed.

Throughout 1893 the inventor was at pains to stress the positive results of his trials, contributing an article to *Engineering* and giving an address to the Society of Arts, but as the months went by and there was no sign of any flying, comment in the press became outspoken to the point of being derisive. By the spring of the new year the Maxim Nordenfelt board was growing increasingly impatient. Although

---

9. With whom Hiram was on friendly terms despite their being on opposite sides during the cordite trial. Moulton found the inventor 'a most engaging personality', and was invited by him to see both the gun and the flying machine in action.

Hiram was sustaining the enterprise largely from his own pocket, costs were escalating and several directors led by Sigmund Loewe issued what was in effect an ultimatum – put the machine to the test of flight or discontinue the experiments. Accordingly Hiram resolved to increase the total lifting surface to 4000 square feet by extending the central section into the form of a biplane, and to make the first serious attempt to get the machine off the ground.

It was a golden opportunity for him to indulge his instincts as a showman, and he took full advantage of it. On 28 July 1894 the inventor laid on a demonstration and reception for the company's employees as a preliminary to the first attempt to 'fly'. The *Kentish Times* reported:

> Mr Maxim is nothing if not original, and the manner in which he received the staff of the Maxim Nordenfelt Gun Company, with their wives and lady friends, at his pleasant residence at Bexley is not the least notable of his doings. That his guests were hospitably entertained was what everybody expected, and a considerable amount of entertainment was derived from the trip with the flying machine. The possibilities of this monster aerial traveller were presented to the minds of the visitors in a very practical manner, and many . . . expressed the hope that when first used in military affairs, it might be employed on the side of, and not against, our own country.[10]

Finally, the press was invited to attend the inaugural 'flight' on Tuesday, 31 July, when a number of reporters and other observers made the trip to Bexley Heath, all amazed by the spectacle of the huge biplane steaming gently at the start of its artificial runway. It was indeed an awe inspiring sight. According to a later article in *Engineering*, the machine was motivated by 'two gigantic propellers, each weighing 135lb and each of them 17'10" in diameter and with blades 5'2" wide . . . each was driven by a compact steam engine poised about eight feet from the ground, and it was estimated that with 375 revolutions a minute a speed of forty mph would be obtained.' The machine was provided with a central plane or wing 50 feet wide plus two small planes 27 feet in length attached on either side, giving an overall span of 104 feet. Despite its size the total weight of the aeroplane, laden with water, fuel, the engineer and two passengers, was no more than 8000lb.

At last the time had come to justify the years of hard work. Conditions were ideal, with little or no wind. Photographers were stationed

---

10. Quoted in James E Hamilton, *The Chronic Inventor*, p48.

at strategic points, and the spectators kept at a safe distance. Hiram himself took charge of the throttles and Tom Jackson and Arthur Guthrie the other controls, and the first run was made with a modest steam pressure of 150lb per square inch, when none of the wheels left the track. The second run, at 240lb pressure, resulted in the machine vibrating to and fro between the upper and lower tracks. Amid mounting excitement, preparations were then made for a third attempt. The machine was attached to the dynamometer and the engines run up to 310lb pressure, registering a screw thrust of 2100lb before the restraining ropes were released.

What followed is best described in Hiram's own words:

> We had not run more than 250 feet when all the weight was lifted off the lower steel track and all four of the small wheels were running on the underneath side of the upper track. After running about 1000 feet the lifting effect became so great that the axle-tree of one of the wheels for keeping the machine down was doubled up. On the breaking of the axle-tree the lifting effect on the other side of the machine became so great that the Georgia pine plank was broken in two and raised in the air, and I found myself floating in the air with the feeling of being in a boat; but unfortunately a piece of the broken plank struck one of the screws and smashed it. I instantly shut off steam and the machine came to a state of rest on the earth, the wheels cutting deep into the ground and leaving no track, thus showing that they had settled down vertically and had not run along on the ground before settling.

In the crash the starboard lower wing was destroyed, one of the propellers damaged and the platform twisted and fractured, but from the markings on the wheels and track it was evident that the massive machine had 'flown' for some 600 feet, the last part being free of restraint. Fortunately no one was seriously injured. Hiram was reported as holding fast to one of the struts and walking away without a scratch, though his two assistants were sent tumbling and Tom Jackson was thrown out on his head and quite badly bruised. It was as well that the machine had not lifted into free flight, for this was not yet its purpose, and neither it nor Hiram was prepared for any such eventuality. In any case, as the inventor was aware, the weight of water consumed by the steam engines was so great that it would have been impossible to remain very long in the air.

Ever afterwards Hiram was to insist that this was the first occasion on which a flying machine had raised itself and a man into the air. In the light of Ader's earlier feat this is questionable, though it must be said that the Frenchman, too, was given to exaggerated claims. What is indisputable is that the episode provided a strong impetus to the debate on the future of manned flight. On 3 August *The Times* reported:

> Mr Maxim's . . . efforts have now been crowned with success . . . it really did rise from the rails . . . After such an experiment few engin-eers will deny the possibility of constructing an aerial vessel so powerful and yet so light as to be able to propel itself and its crew through the air, together with water and fuel sufficient for a voyage.

Other reactions were mixed, the *Illustrated London News* noting that the accounts of what took place 'are a little various, and suggest dif-ferent degrees of refreshment in the reporters . . .'. The *Pall Mall Budget* waxed lyrical, announcing that 'Mr Hiram Maxim has suc-ceeded where Daedalus failed. He has solved the problem of aerial navigation, and has mounted in the region of the air, if not upon "the wings of the dove", at least upon the broad pennons of his aeroplane.' But after due consideration the periodical *Engineering* was more guarded:

> It is not to be supposed that the machine approaches anything like perfection, but there can be no doubt that its performances consider-ably eclipse previous efforts . . . Maxim has produced a very curious and interesting machine, but he is not one step nearer to flying than any of his predecessors . . . he has made a machine that does not know how to fly . . . We believe that the safe use of such a machine will always depend on the skill of the driver. Because we put on skates, we do not expect be able to skate straight away. No more can we expect Mr Maxim to be able to control this machine efficiently until he has had years of practice.

In an article for the *American Engineer and Railroad Journal* Octave Chanute was similarly cautious: 'Mr Maxim's aerial enterprise has . . . not failed ignominiously; its fall, indeed, may be described as brilliant, since success was the immediate cause of the disaster that followed.' However, he asked, what has it taught us? On the essential issue of maintaining equilibrium in the air 'as it involves ascent, flight and descent, the answer is, absolutely nothing'. Moreover the machine itself was so 'vast, unwieldy and . . . of the most delicate fragility', that it was unlikely to be able to support itself in the air. 'It is my private opinion', he wrote to one of his correspondents, 'that Mr Maxim will now build an entirely new aeroplane, as the stability of the present one is evidently unreliable, and it would be a mistake to repair it. His great achievement is the steam engine, and if he applies it to a new aeroplane about half the size, I deem it not impossible that he may show some remarkable results . . .'[11]

Already Chanute, who was to design and build his own multiplane and biplane gliders and go on to motivate and inspire the Wright brothers, was tending to diverge from Hiram and Langley in his approach to aeronautics. He probably also disapproved of his fellow countryman's preoccupation with the military aspects of the subject. Something of an idealist himself, he preferred to believe that 'the advent of a successful flying machine . . . will bring nothing but good into the world; that it shall abridge distance, make all parts of the globe accessible, bring men into closer relation with each other, advance civilisation, and hasten the promised era in which there shall be nothing but peace and goodwill among men.' In his classic work *Progress in Flying Machines*, however, he was in no doubt that the name of Maxim 'must ever remain as that of one of the men who have hitherto done the most to advance the solution of the problem of aviation.'

Hiram himself declared in a letter to *The Times* that he had shown 'that it is possible for a machine to be made so light, and at the same time so powerful, that it will lift not only its own weight but a considerable amount besides, with no energy except that derived from its own engines . . . There can be no question but that a flying machine is now possible without the aid of a balloon.' To a correspondent who accused the inventor of endangering his own and other peoples' lives and predicted catastrophe if he persisted in his experiments, Hiram replied conceding that flying did indeed involve risks. But he also pointed out, underlining once again the motivation of his researches, that all weapons of war were dangerous – only more dangerous was to be on the receiving end:

> In view of the decided advantage which a flying machine would give its possessor over an enemy, I do not think that in case of war European nations would hesitate to employ them even if one half of the men navigating them were killed . . . War, at best, is a dangerous game, and those entering upon it are playing with dangerous instruments, whether they are guns, dynamite or flying machines . . .

In a further letter to *The Times*, Lieut B [F S] Baden-Powell, Scots Guards, weighed in on Hiram's side. Brother of the later hero of

11. Chanute to Thomas Moy of the Patent Office, London, 17 August 1894. Chanute Papers.

Mafeking and Chief Scout, he was soon to become secretary of the Aeronautical Society and was himself engaged at the army's Balloon Factory at Aldershot in experiments with man-lifting kites for purposes of military observation. 'No good reason,' he wrote, 'can, I believe, be adduced why man should not construct an efficient flying apparatus which, when accomplished, would rank among the greatest inventions of the age, and it seems a great pity to deprecate a genuine attempt at its solution, when no other reason can be brought against it except that it is presumed to be dangerous to the inventors.'

On 10 August a paper prepared by Hiram on 'The Evolution of a Flying Machine' was read by Brodrick Cloete at a meeting of the British Association at Oxford. The gun company was naturally anxious to give the widest possible publicity to Hiram's achievement in the hope that this would lead the War Office to subsidise further research, and the *Illustrated London News* duly reported that his paper 'furnishes trustworthy information that we are within measurable distance of the faculty of flying, [this being] acknowledged by no less authorities than Lords Kelvin and Rayleigh, the latter of whom remarked that the inspection of Mr Maxim's flying machine was one of the sensations of his life . . .' It also went on to reveal that:

> Mr Maxim has had for some time another flying machine on the stocks, which will correct . . . all previously detected errors. According to Mr Maxim's exact calculations, the expenses up to date of his flying machine amount to £16,935.7.3d . . . its capabilities as an engine of war are so invincible that fortifications, navies and armies would become mere details . . . as to the relative superiority of nations in respect of armament.

Doubts, however, persisted with regard to the Baldwyn's Park demonstration. To many of the onlookers the upward movement of the machine had seemed barely perceptible, and they went away disappointed, having expected to see it rise up and disappear over the treetops in a manner familiar from artists' impressions in the popular papers or the writings of Jules Verne. Luckily H G Wells, who was fascinated by the whole episode and had spoken and corresponded with Hiram, was on hand to supply the deficiency. In his short story *The Argonauts of the Air*, which appeared in 1895, Wells provided an alternative and more sensational version of what he thought *might* have happened. This is of more than passing interest since it obviously reflects the experience of the inventor and the views of the two men concerning the practicalities of flying.

Monson, the hero of the story, had we are told: '. . . taken up the work [of building a flying machine] where Maxim had left it, had gone on at first with an utter contempt for the journalistic wit and ignorance that had irritated and hampered his predecessor, and had spent . . . rather more than half his immense fortune upon his experiments . . . It [Monson's machine] was a great and forcible thing beyond dispute, and excellent for conversation; yet, all the same, it was but flying in leading-strings, and most of those who witnessed it scarcely counted its flight as flying. More of a switchback it seemed to the run of the folk . . . and if [the inventor] did not mind the initial ridicule and scepticism, he felt the growing neglect as the months went by and the money dribbled away . . .' Monson's idea was 'to get into the air with the initial rush of the apparatus', as does a rook or gull, but the fact was that 'the bird is practising this art from the moment it leaves its nest'. A man has difficulty even in balancing on a bicycle, and 'the instantaneous adjustments of the wings, the quick response to a passing breeze, the swift recovery of equilibrium . . . all that he must learn with infinite labour and infinite danger, if ever he is to conquer flying.'

In the story Monson and his foreman do succeed in rising up and flying gloriously across London before they lose control of the machine and it crashes in flames. But the record of the inventor's work remains 'to guide the next of that band of gallant experimentalists who will sooner or later master this great problem of flying.' The distinction made by Wells between the effort involved in taking off and the skills required to control the machine in the air was of course a crucial one. It had already been emphasised by Chanute and by the German Otto Lilienthal, who was succeeding with ever more ambitious flights in his hang-gliders and was critical of Hiram's efforts, declaring that the American was on the wrong track and that even if his machine had taken off he could not have maintained it in flight. This was true, and indeed Hiram was the first to admit that much practice would be needed before he could expect to achieve such control. For his part he affected to dismiss Lilienthal as a mere parachutist, likening him to a 'flying squirrel', and the German responded in kind, retorting that the one achievement of Hiram's massive aeroplane had been to show others how not to fly.

Undeterred by the derailment on 31 July and repudiating any suggestion that the experiment had been a failure, Hiram at once set about repairing the damage to his machine, fitting stronger outriggers and replacing sections of the wings and the buckled platform. But although he had every intention of continuing the work, the money was fast running out. Basil Zaharoff was later to claim that in return for helping to finance Hiram's flying experiments he had been given a trial 'flight', and so was one of the first men to be lifted off the ground in a heavier-than-air machine. He was, however, unwilling to do more, and as the months went by it became clear that the British authorities were not going to follow the example of the French in providing subsidies for Clément Ader. Given the parlous state of the Maxim Nordenfelt company's finances, even Albert Vickers and Hiram's other supporters on the board, discouraged by the lack of any immediate success, felt unable to press for continued funding.

The inventor therefore decided to appeal directly to the public and augment his dwindling funds by charging for admission to a series of exhibitions of his flying machine and automatic guns at Baldwyn's Park. The first of these, mounted in aid of the Bexley Cottage Hospital, took place on a damp day in November, when the *Bexley Heath Observer* reported that 'a crowd of 2000 persons willingly paid their florins' to witness the machine in action, even though it was 'divested of several of its aeroplanes' and was not intended to fly. After Hiram had given his address the engines were started, and as the Bexley Temperance Brass Band played and the propellers went round 'people clutched their hats, and an umbrella was turned inside out, and then the machine shot on its course'. Afterwards the inventor startled the spectators by firing off several belts of cartridges from his gun until, as the rain began to fall more heavily, the proceedings had to be brought to a close.

It is not recorded how successful this exercise was financially, but it was not repeated, and shortly afterwards the final blow to Hiram's aspirations came when the London County Council announced that they intended to purchase Baldwyn's Park for conversion to a mental asylum.[12] The inventor was quick to note the irony. 'It appears,' he commented wryly, 'that I had prepared the ground, so that all that was necessary was to erect the buildings.'

In August 1894 Mr Bucknall Smith, he who earlier had ridden on the flying machine, interviewed Hiram and Sarah in their mansion at Baldwyn's Park for the already quoted article in *The Strand Magazine*, which sheds an interesting light on their situation at the time. At one side of the house was 'an extensive wooden structure from which proceeds a broad-gauge railway – the inventor's workshop and the home of his famous flying machine.' On entering the vestibule, Bucknall Smith discovered, 'an elaborately finished and mounted Maxim gun meets the gaze of the visitor.' The house itself, while avoiding

---

12. In later years the Bexley Mental Hospital, now the Bexley Hospital.

anything in the nature of ostentatious luxury, was 'handsomely furnished throughout in becoming taste', though most of Mr and Mrs
Maxim's leisure hours were spent in the library, 'which is well equipped with most of the leading modern text-books and works of
reference.'

The reporter had more than one session with Hiram, whom he found
to be: '. . . a man of diligent study, great vitality, and resolution. Knotty
points and difficulties only stimulate his unremitting labours and
dogged persistence . . . As a companion he is highly entertaining and
genial, with a distinctly humorous bent . . . he is of medium height, but
powerfully built, erect in gait and agile; his hair is now of silvery hue,
but physically and mentally he is a well-preserved man. His dark-brown
penetrating eyes are full of intelligence and vivacity – in short, his
presence is unusually commanding . . . at times he speaks with slow and
thoughtful emphasis, at others, with the volubility remindful of his gun.
His accent is practically that of an Englishman, although most of his
quaint idioms are decidedly American.' As to the inventor's 'intelligent
and devoted wife', she is a 'very industrious, cultured and genial lady,
who not only takes immediate interest in all her husband's inventions
and scientific pursuits, but directly assists him in various departments of
his daily avocations; indeed, their labours and deliberations may be
appropriately described as inseparable.'

Bucknall Smith reminded his readers that Hiram had devoted only
his leisure time to the problem of mechanical aerial flight, which
might therefore be seen as one of his hobbies. Asked about the prospects following his flying machine experiments, the inventor was cautious: 'I think too much has been already said and written about this
machine, and much nonsense embodied, although I thought it desir-
able that my achievements in this direction should be recorded, so that
if anything happens to prevent me from proceeding with my experiments, someone else may take them up and bring them to a practical
issue.' In fact Hiram was well aware of the limitations of his machine
and in particular the problem presented by his inability to 'steer' it in
the air. Throughout most of 1895, while busy with other matters, he
worked intermittently on the design of a more efficient power unit,
but otherwise he rested on his laurels while considering how best to
raise money for a further step forward.

One way of doing this was to engage the interest of those most
likely to be able to help. In April the Duke of York (later George V),
accompanied by Admiral Commerell, paid a visit to the Erith factory
and also to Baldwyn's Park, where a run on the machine was arranged
for his benefit. On this occasion, as Hiram recorded with some glee:

... we did not let go until there was a screw thrust of 2000lb; of course the machine bounded forward with very great rapidity. Admiral Commerell became frightened and said: 'Slow up,' but the Prince [*sic*] retorted, 'Let her go for all she's worth', and I did. The Admiral was greatly frightened when he found that we were going at railway speed with the woods only 200 feet away, but the three strong ropes and the rotating windlasses very soon brought us to a state of rest.

In August the journal *Nature* reported that a few weeks earlier 'a large party of scientific men' had made the journey to Baldwyn's Park at the invitation of Hiram and Cloete to inspect the repaired machine. Now that the main mechanical difficulties of construction had been overcome, it noted, a longer track was required for the purpose of practice in vertical steering while the machine was off the ground, but bearing upwards against the outer rails. 'It is unfortunate that difficulties should have been thrown in the way of making an extension of the present track . . . so another practice ground, perhaps a sheet of water, must be found, not too far from headquarters or from skilled assistance.' But this was wishful thinking. Already Hiram was under notice to quit Baldwyn's Park by the end of the year, and the only question was where else the experiments might be carried on.

It was just as the American was pondering the question of how to manoeuvre his machine in the air that he came into contact with Percy Sinclair Pilcher, that most venturesome of early British aeronauts. An assistant lecturer in the department of naval architecture and marine engineering at Glasgow University, Pilcher had for some time, with the aid of his sister Ella, been building wood and fabric hang-gliders of his own design and flying them on the Dumbartonshire hills. In this he followed in the footsteps of Otto Lilienthal, whom he visited in Germany and like whom he had been conducting increasingly successful 'soaring' experiments, running downhill and taking off to travel ever longer distances before landing. By the autumn of 1895 Pilcher was sufficiently confident about his ability to control his glider in flight that he applied to the university's professor of natural philosophy, the distinguished physicist Lord Kelvin, for financial assistance to enable him to construct a compact engine, possibly powered by carbonic acid gas, and so extend his range.

Lord Kelvin had, it seems, visited Baldwyn's Park on more than one occasion, paying Hiram a somewhat backhanded compliment by remarking that his workshop was 'a perfect museum of invention', but despite this, or perhaps because of it, he took a jaundiced view of what he perceived to be another ill-conceived stunt. A year later he was to decline an invitation to join the Aeronautical Society, writing to its secretary, Major Baden-Powell: 'I was greatly interested in your work with kites but I have not the smallest molecule of faith in aerial naviga-tion other than ballooning or of expectation of good results from any of the trials we hear of . . .' Accordingly he was reported as saying 'that on no account would he help [Pilcher], nor should I, as he should certainly break his neck.'[13]

Percy Pilcher therefore wrote to Hiram to seek his advice. Having followed with interest the American's flying experiments, he suggested that despite their differences in age and experience it might be useful for them to compare notes with a view to collaboration. Hiram was never one to turn down the offer of help from bright young men. To match Pilcher's practical skills with his expertise in the design of powered machines made every sense, and after an exchange of corre-spondence it was agreed that he should travel south to join Hiram's team. This was an encouraging development, as was the decision of the Maxim Nordenfelt board to allow Hiram to transfer his workshop and hangar from Baldwyn's Park to the company's firing range at Eynsford in Kent. It was not, however, prepared to provide any more money, and so, to defray costs, Hiram invited his friends and suppor-ters to subscribe to a new syndicate floated to keep the enterprise alive.

In January 1896 Hiram and Sarah vacated Baldwyn's Park, moving back to Thurlow Lodge at Norwood. In March Percy Pilcher, having resigned his post at Glasgow University, went to work as Hiram's assistant at what was styled the Maxim Nordenfelt 'Experimental De-partment', taking with him his *Hawk* and *Gull* gliders. Together the two men supervised the relocation of the workshop and hangar at Upper Austin Lodge, not far from the Eynsford range, where at once they began to prepare for an International Exhibition of Motors and their Appliances held in May at the Imperial Institute in South Ken-sington. This featured one section devoted to 'Apparatus for Aerial Navigation', which contained a model of Maxim's flying machine and one of his steam engines as well as a Lilienthal glider and Pilcher's new *Hawk* glider.

Owing to Hiram's other preoccupations, most of the work had to be done by Pilcher, who in June went again to Germany to see Lilienthal fly his latest glider and discuss with him possible means of achieving extended flight. On this the German had no very definite views, al-though he clung to the principle of the ornithopter, experimenting

13. Philip Jarrett, *Another Icarus, Percy Pilcher and the Quest for Flight* (Washington DC 1987), p29.

with mechanical 'flappers' while concentrating on the furtherance of his skills as a hang-glider pilot. 'Dexterity alone,' he maintained, 'invests the native inhabitants of the air with superiority over man in that element.' But to Pilcher, as to Hiram, hang gliding was only a step towards powered flight, and he wrote to a fellow enthusiast in Australia, Lawrence Hargrave: 'To my new machine [the *Hawk*] I hope shortly to be able to add, with Mr Maxim's help, a small oil engine and a screw propeller, and then starting from an eminence I shall be able to make attempts at horizontal flight.'

In May 1896 news came from Washington that S P Langley's model *Aerodrome No. 5*, weighing 30lb, had flown for nearly two minutes, travelling a distance of 3300 feet in a series of graceful curves before descending into the Potomac River. Langley, delighted, announced that 'the great universal highway overhead is very soon to be opened', and Hiram agreed, interpreting this latest feat as confirmation that a heavier-than-air machine could take off and steer itself with a minimum of human assistance. He wrote to *The Times*: 'Professor Langley has made the working model, and . . . I feel sure that it is now possible to make a successful and practical flying machine which will at least be a valuable adjunct to the offensive and defensive powers of highly-civilised nations who are able to make and operate delicate and complicated machinery.' He was, however, having second thoughts about the configuration of his own machine, adding in the same letter: 'If machines are to be made on the aeroplane system, small ones will be found to work much better than large ones.' Under the influence of Langley and Chanute he was coming to believe that he had been over-ambitious: 'Instead of making such a large machine, I should have experimented with a much smaller one, and been sure of my practice ground before commencing experiments.'

At a time when unencumbered, conveniently located open spaces were not easy to find, this last was a factor of major importance. The first problem facing any would-be aviator was how to get himself into the air; the second was how to get down again without damage or injury. Hiram saw this as the main obstacle to further progress for which, he wrote, 'it will be necessary to obtain a very large and level field completely free from trees and houses . . . I do not consider it safe to attempt free flight with . . . very large trees in every direction. What is required is to experiment with the machine running very near the ground . . . [and] not until one has complete control of it should high or completely free flight be attempted.'[14] In this connection he

14. *Century Magazine* (February 1895).

approved of Langley's technique of carrying out his experiments over water, so hopefully minimising the damage caused to his machines and enabling regular adjustments to be made.

Percy Pilcher employed a towing technique which enabled him to make some excellent glides in the *Hawk*, the longest of some 750 feet, but for any more ambitious machine Eynsford offered little scope. Characteristically, therefore, Hiram reverted to the idea of a helicopter. During the next twelve months he and Pilcher explored the possibility of combining the vertical lift this would offer with the power generated by a lightweight internal combustion engine of the type being developed for automobiles. In the spring of 1897 Pilcher informed a meeting of the Aeronautical Society: 'Since the accident to his big machine Mr Maxim has done very little. I am working with him, and I believe that we are shortly going to begin a machine of quite new type about which I am not at liberty to speak.' Soon afterwards Hiram filed a new patent for a large twin-rotored helicopter powered by 4-cylinder engines, but this was destined never to see the light of day.

Sadly, events were conspiring to ensure that the partnership between Hiram and Pilcher was to have no positive outcome. In August 1896 the doubts of those who believed all attempts at flying to be foolhardy were confirmed when Otto Lilienthal crashed with his hang-glider and was killed. At Eynsford Pilcher continued to make flights of up to a hundred yards in the *Hawk*, at the same time giving lectures and appealing for financial support, but the response was meagre. The Maxim Nordenfelt board had little sympathy either with Hiram's helicopter experiments or his efforts to set up independent syndicates to finance his research into aero and automobile engines. In any case, plans were already well advanced for the amalgamation of Maxim Nordenfelt with Vickers, an arrangement which was completed in September 1897. Although the new concern was called Vickers, Sons and Maxim, reflecting the importance of Hiram's role in its development, Sigmund Loewe remained determined to make its operations more cost-efficient. He therefore proceeded to curb what he saw as the inventor's more unproductive enthusiasms, and as a step to this end he required Hiram to remove his workshop and hangar from the vicinity of the Eynsford range.

This effectively brought to an end the joint endeavours of Maxim and Pilcher, who was left with no alternative but to part company with his mentor. It is at least possible, for they were moving on the right lines, that had the two men been able to continue they might have anticipated the Wright brothers' epoch-making success of five years later. But it was not to be. Hiram had become disheartened by the repeated failures and the apparently limitless expense involved in flying experiments.[15] During the next two years Pilcher tried to develop a lightweight aero-engine of his own, setting up his own company and corresponding first with Lawrence Hargrave in Australia and then with Octave Chanute in America on the most suitable design of aeroplane in which to fit it. But in October 1899, while demonstrating his classic glider the *Hawk*, Pilcher, aged 32, suffered fatal injuries when the machine collapsed in flight and plunged to the ground.

'Hindsight,' writes Pilcher's biographer, 'enables us to see that [his] decision to apply power before fully mastering control and the Wright [brothers'] mastery of control before the application of power were, respectively, the routes to failure and to ultimate success.' But to contemporaries Pilcher's death was just another deterrent to experiments with flying machines, confirming the worst fears of Lord Kelvin and those who thought like him. With the arrival of the new century the prospects for a practical powered aeroplane seemed as remote as ever. In Europe the focus of aeronautical interest switched back to balloons and dirigible airships driven by automobile-type petrol engines, and the initiative in respect of heavier-than-air machines passed to the United States.

Although Hiram no longer had the means to conduct flying experiments, his interest in the subject was only partly abated and his reputation as a pioneer aeronaut stood high. He could not help feeling a certain bitterness when in 1898, under the stimulus of the Spanish-American War, the United States War Department at last recognised the military potential of manned aircraft and provided S P Langley with $50,000 to finance his further experiments. These resulted in a series of trials with ever larger flying machines culminating in the full-size, piloted *Aerodrome A*, which in December 1903 was launched by catapult from the roof of a houseboat on the Potomac, only to plunge straight into the river. Once again it was shown that an aeroplane, however well engined and stable in design, could not be driven into the air and expected to stay there with the navigator making adjustments as to a ship at sea.

In that same month, near Kitty Hawk, North Carolina, and without the benefit of government subventions, the Wright brothers proved that they had found the right formula: first master control in the air, and only then add the power. The secret of the Wright *Flyers* was that

---

15. In October 1897 Clément Ader's military aeroplane was wrecked while making its first flight, and the French authorities refused to advance money for further experiments.

they were deliberately constructed to be unstable and directed by an experienced hand. 'What is chiefly needed,' wrote Wilbur in 1900, echoing the philosophy of Lilienthal, 'is skill rather than machinery.' The first epic flight on 17 December 1903 lasted only twelve seconds, but the fourth and final flight that day lasted 59 seconds and covered 852 feet. As Orville pointed out, it was the first occasion on which 'a machine carrying a man had raised itself by its own power into the air in full flight, had sailed forward without reduction in speed, and had finally landed at a point as high as that from which it started.' It was the beginning of a new era, although this was not recognised at the time or for some years afterwards.

The rightful place of Hiram's flying machine experiments in the history of aviation remains a matter of dispute. Few contemporaries doubted that his contribution was both original and significant. The charismatic inventor continued to be in demand as a speaker and writer on aeronautical matters, and his *Artificial and Natural Flight* was received with respect, despite some lukewarm reviews. Thus the journal *Autocar* was inclined to dismiss it as 'a popular description of work that has been already accomplished . . . which cannot be compared with the new classical volumes of F W Lanchester, who has approached the work . . . in an altogether different spirit.' Others were put off by Hiram's typically exaggerated claim that all later machines were based on his 1894 model:

> . . . all have superposed aeroplanes of great length . . . all have fore and aft horizontal rudders, and all are driven by by screw propellers . . . in this respect they do not differ from the large machine that I made at Baldwyn's Park . . . and the fact that practically no essential departure has been made from my original lines, indicates to my mind that I had reasoned out the best type of a machine even before I commenced a stroke of the work.

More recently Harald Penrose, himself an aviator and test pilot of distinction, has in his *British Aviation, the Pioneer Years, 1903-14*, endorsed the importance of Hiram's achievement, highlighting the quality of his experimental work with lightweight steam engines, aeroplane wings tested in a large wind tunnel and propellers of laminated wood. On the other hand Charles Gibbs-Smith's comprehensive *Aviation, An historical survey* is as contemptuous of the American's efforts as of the claims he makes in his book, which he describes as 'egregious nonsense'. In general Gibbs-Smith adopts a distinctly magisterial tone, dividing the early pioneers into two categories, the 'chauffeurs', who misguidedly saw the flying machine as a 'winged automobile, to be driven into the air by brute force . . . and sedately steered about the sky', and the 'true airmen', who had the good sense to think primarily in terms of control in the air and rode their machines 'like an expert horseman'. If, he avers, the glider approach to flying is not followed, 'the risks and dangers are immense; and it was foolhardiness or ignorance which prompted those early pioneers who attempted to build powered aeroplanes to start with.' Since Hiram was 'chauffeur-minded', his experiments were a waste of time and money, his contribution to aviation was virtually nil and he influenced nobody.

In effect Gibbs-Smith condemns the American for setting himself too limited an objective. Hiram's biplane, he correctly points out, was 'not in any true sense a flying machine, but what one might now call a huge "lift test rig" '. The inventor was, moreover, 'not in the least interested in the practical problems of flying: he was concerned with lift and thrust, and demonstrated their workings; but that was all, and it was in no way original.' But while in retrospect this is seen to be true, it seems a little hard to dismiss men like Ader, Maxim and Langley for not perceiving what was far from obvious at the time they were working. Leaving aside the fact that when he embarked on his experiments no automobile was yet in existence, it must be borne in mind that Hiram started virtually from scratch. There was no consensus on what a flying machine should look like, no aero engine, no agreement on the configuration of wings or propellers, and no means of lifting the machine off the ground. Every detail had to be worked out by a process of trial and error, and given the limitations on the inventor's time and resources it is remarkable that he was able to accomplish as much as he did. In the words of his erstwhile collaborator Edward Hewitt: 'Hiram Maxim was the first scientific experimenter on the aeroplane and the first to obtain aeronautical data of any accuracy and value for flight. He should be accorded far more credit than is now given him for his pioneering work.'

As will be seen, Hiram's influence on aeronautics did not end with the abandonment of his experiments by the Vickers company, which in due course was to follow up his initiative in aircraft design as it had followed up his work on the automatic gun. The American continued to take an active interest in the advances made during the early years of the new century, and to be consulted by a new generation of pioneer aviators, those 'magnificent men' who were to set the seal on practical flying before 1914. Notable among these was Claude Grahame-White, the first Englishman to hold an official pilot's licence, who was to work briefly with Hiram and whose considered judgement on the inventor's experiment of 1894 is perhaps as sound as any. 'Although interesting and ingenious,' he wrote in 1912, 'no very practical results were achieved with this giant machine, but it yielded a great amount of data, and so served its purpose.'[16]

---

16. C Grahame-White, *Aviation*, p27.

FIVE

The Parting of the Ways

I treated Hiram with almost reverential deference, and I would have worked my fingernails off for him. To be sure, I wished to make money, but my chief aim in life was to do something worth while . . . and I couldn't keep all my abilities submerged. So we vibrated back and forth more or less between friendship and open enmity [until] finally the inevitable break came.'

HUDSON MAXIM

Following the creation of Vickers, Sons and Maxim in Queen Victoria's Jubilee year of 1897, an ambitious programme of expansion was begun in the company's shipyards, steel mills and gun factories, the plant at Erith, for example, being expanded to cover eighteen acres and to employ some 4000 workers. Heavy naval guns and their massive mountings were produced at Sheffield and Barrow, automatic guns and medium and quick-firing artillery at Erith and Crayford, and both the rifle calibre and 37mm 1-pounder Maxims stimulated a rising volume of orders from all over the world.

Hudson and Lilian were still living in England, but while the brothers went about their business at least outwardly on good terms, there was growing friction between them. Hiram persisted in rubbing Hudson up the wrong way by adopting a patronising attitude towards his separate initiatives and insisting on calling him 'Ike'. Hudson for his part annoyed Hiram by registering a steady stream of American patents arising from the inventions he claimed to have made in consultation with ordnance and explosives experts in the United States. This he did not with any deliberate intention of offending his brother, but as he put it somewhat disingenuously:

Hiram was exceedingly jealous of his own fame, and couldn't tolerate a rival . . . he seemed to have a sort of mental twist about other inventors stealing his inventions. He told me one time that if the telescope hadn't been invented he would have invented it; and I think he never felt kindly toward Galileo for having got ahead of him. My own inventions irritated him to the last degree.

For the moment, however, each was too busy pursuing his own concerns to pay much attention to the activities of the other. Hudson and Lilian continued to reside at Thurlow Lodge, where they made a modest contribution to the rent. Hiram and Sarah spent much of their time at their town house in Kensington, 18 Queen's Gate Place, close to the Science Museum, from where the inventor travelled daily to the gun factories at Erith and Crayford or to his Experimental Department with its office at Victoria and workshop at the Eynsford range. In the matter of experiments with aeroplanes and helicopters, as with Hiram's endeavours to produce a serviceable automobile, Hudson had only an incidental interest, but as always he was glad to learn from Hiram's experience, and the two men collaborated amicably enough on associated projects such as the auto-car and the carbide and acetylene gas syndicates.

From every direction, however, the storm clouds were gathering. Hudson's main concern was to further the cause of his Maxim Powder and Torpedo Company, and to this end he remained in close touch with the New York office and the works in New Jersey while corresponding with the American naval and military authorities on the subject of the smokeless powder, high explosive projectiles, torpedoes and other inventions which increasingly he presented as his own. The most urgent priority, and the reason why he was spending so long in England, was to persuade Vickers or some other established explosives company to undertake the manufacture of the Maxim-Schupphaus smokeless powder for sale in Europe. In fact, as it happens, this was a wild goose chase, for unbeknown to Hudson developments were taking place behind the scenes which made any such eventuality unlikely.

In the spring of 1897 Sigmund Loewe, together with other representatives of the Anglo-German group of explosives companies, travelled to New York to meet their counterparts in the American group. Their purpose was to renew an earlier profit-sharing agreement made in 1888 by which the companies divided up the world market between them.[1] This was a common enough practice, similar arrangements already being in existence to control the marketing of other commodities such as coal, steel, armour plate and non-ferrous metals. Broadly speaking, the British and Germans undertook to refrain from doing business in the United States and the Americans undertook to keep out of Europe, Africa and much of Asia, both groups having equal rights in Central and South America, China and Japan. This

---

1. Details of these agreements are given in A D Chandler and Stephen Salsbury, *Pierre S du Pont and the making of the modern corporation* (New York 1971), pp171 *et seq*.

meant that the American armed services had to pay a royalty on all military explosives and propellants produced by the European companies, and vice versa, and so it was clearly not in the interest of one group of companies to buy explosives patented in a country in the other group.

Since at this time negotiations were already well advanced to bring about the merger of Maxim Nordenfelt with Vickers, Sigmund Loewe went to the States as one of the more influential delegates of the Anglo-German group. The new agreement having been ratified by Du Ponts and other leading American firms and accepted in principle by the US Bureau of Ordnance, he returned to London to report to the Maxim Nordenfelt directors, who received his account of the proceedings with satisfaction and rewarded him with 'a hearty vote of thanks'.[2] So far, since the Americans were not yet in a position to produce the latest high explosives and smokeless propellant for themselves, the arrangement had worked to the advantage of the European side, while the Maxim Nordenfelt board saw in it a means of removing the uncertainty surrounding the powders developed by the Maxim brothers, especially those patented by Hudson in the United States.

By the summer of 1897 Hudson was in something of a quandary. None of his approaches to armament companies in England or on the Continent had met with a positive response, and he and Lilian were embarrassingly short of money, having to draw on her meagre savings. For some weeks the couple lived on credit, being down to their last few pounds in cash, and in desperation Hudson cast around for alternative sources of income. In May he published an article in the *Pall Mall Gazette* in which he claimed to have found a means of producing artificial diamonds, and then, drawing on discussions with Hiram and Hiram Percy, he wrote to potential manufacturers offering to sell his ideas on a 'process of making tubes from solid ingots' and a 'motor for road carriages' fuelled by benzine or naphtha vapour. This motor, he explained in a circular letter, 'shall be light and compact, and still capable of developing sufficient energy to propel a carriage at any desired speed, even when running up hill, silent in its operations and perfectly under the control of the driver . . .'.

But although none of these initiatives was to bear fruit, events were beginning to move in his favour. In March Hiram and the Maxim Nordenfelt company were disappointed when in the court action they had brought to establish the priority of his smokeless propellant over cordite (referred to in the legal record as 'Maxim-Nordenfelt versus

---

2. Maxim Nordenfelt minute book, April 1897.

Anderson and the British Government'[3]), judgement was finally entered against them. 'No one,' reflected Hiram ruefully, 'ever wins against the Government', and indeed his barrister friend Lord Moulton, who on this occasion had been briefed by the company, believed that Hiram had been defeated on a technicality in that he had used castor oil in his powder rather than a true mineral oil to reduce the temperature of the explosive gases: 'But for this I think he would have succeeded in the long litigation which he had with the Government . . .'[4]

To Hudson the decision marked something of a watershed. Hitherto he had been inhibited by the knowledge that he and his brother had worked together on propellants of broadly similar composition, but now that Hiram's British patent had been formally excluded from the public domain he felt free to publicise the virtues and advantages of the Maxim-Schupphaus powder which he claimed to have developed and patented independently in the United States.

The situation further changed in June, when, possibly as an indirect result of the cartel agreement reached in New York, Du Ponts informed Hudson that they wished to exercise their option, established two years earlier, to buy up his Powder and Torpedo Company together with all his inventions and the American patents for the Maxim-Schupphaus smokeless propellant. Their terms were not exactly generous, a total of $175,000 to be paid in annual instalments of $25,000 plus a commission on powder sales, but Hudson was in no position to bargain. Having consulted his backers and shareholders in New York, he agreed to the deal. After the company's creditors had been paid off, he and Dr Schupphaus each received $15,000, which in Hudson's case was hardly sufficient to cover his debts. Still, it was a step in the right direction, and if a market could be found for his explosives and other inventions as manufactured by Du Ponts, money was undoubtedly there to be made. The problem was that such demand as there was for munitions existed in Europe rather than America, and so Hudson redoubled his efforts to find buyers in England.

Here as in the rest of Europe the armament industry was concentrating on naval expansion and particularly the building of ever more powerful battleships. These were in effect floating batteries of heavy guns, and the thrusting new Vickers company was mainly concerned to challenge Armstrongs (soon to combine with their old rival Whitworth) as the leading supplier of naval ordnance to the

---

3. Sir William Anderson was Director-General of Ordnance Factories at Woolwich.
4. Mottelay, Introduction, xxi.

Admiralty. One difficulty common to both was that the new propellant powders were tending to erode the barrels of large naval guns
at an alarming rate. To achieve longer ranges British cordite and, to a
lesser extent, German ballistite contained a high proportion of nitroglycerine which generated hot gases so violent in their effect that the
inner linings of the guns were having to be replaced at ever-shorter
intervals. Limiting as it did the frequency of firing practices, this was a
problem with serious tactical implications which to Hudson seemed to
offer a selling point and a means of breaking into the European
market.

Having become adept at enlisting support by way of the proceedings of learned societies, Hudson arranged to give lectures at the
Society of the Chemical Industry and the Royal United Services
Institution on, respectively, the smokeless cannon powder worked out
'with the valuable aid of my colleague, Robert C Schupphaus, Ph D',
and an aerial torpedo which he claimed as his own. He also had
printed and circulated to interested parties a book entitled *The Maxim
Aerial Torpedo, a New System of throwing High Explosive from Ordnance*.
This was described as a naval projectile 'twice the calibre, double the
length and three times the weight of the present armour-piercing
shell', which was propelled by the Maxim-Schupphaus powder,
charged with guncotton or the provocatively entitled Maximite ('a
high explosive invented by me and made according to my United
States Patent No. 544924'), and detonated by a delayed-action fuse
developed in collaboration with Professor Alger of the US navy. All
these inventions, Hudson declared, were the property of the Maxim
Powder and Torpedo Company, which had succeeded the Zalinski
Pneumatic Dynamite Gun Company of New York and ensured for the
first time that a high explosive shell could be fired with safety from
heavy ordnance.

Hudson followed up his lectures by a succession of articles including
one of July 1897 in *The Engineer* entitled 'Some new features in
smokeless powders and their ballistic results'. This delivered a straight
sales pitch for the Maxim-Schupphaus propellant and the Maxim
aerial torpedo. Because of the risk of premature detonation, it was
argued, 'no pure gun-cotton [*ie* nitrocellulose] powder, except in the
form of the Maxim-Schupphaus multi-perforated grains, has yet been
produced which has been successful in guns of more than 5in calibre'.
Unlike propellants with a high nitroglycerine content, moreover, the
Maxim-Schupphaus powders with 9 per cent or less nitroglycerine did
not erode the bore of heavy guns, while their 'high excellence . . . is
well evidenced by the fact that they are now being manufactured and

furnished to the United States Government by the world-renowned firm of E I Du Pont de Nemours . . . who have acquired the American patent rights.'

The article was accompanied by tables summarising the results of test firings with the Maxim-Schupphaus powder at Sandy Hook, together with a large diagram of 'my improved aerial torpedo and fuse, carrying a ton of wet gun-cotton and designed for a gun of 24in calibre'. This huge projectile, especially when charged with a high explosive such as lyddite or the inventor's own Maximite, was allegedly capable of being fired to a distance of 5 miles and on landing near its target could be expected to 'inflict a fatal injury upon the hull of the strongest battleships now made'. Included by way of contrast was a drawing of the existing Whitehead service torpedo, with its much smaller explosive charge and range of less than 1 mile. Hudson was careful to disassociate the Maxim-Schupphaus powder from the propellants developed by his brother: 'In view of the late Maxim-Nordenfelt suit versus Anderson and the British Government,' he wrote, 'it may be well to mention here that [our] inventions . . . have no connections with those of Mr Hiram S Maxim.'

Hiram's reaction to these initiatives, however, can readily be imagined. Adding to his chagrin at losing out in the case against the Government's cordite patent, it now appeared that Hudson was taking advantage of his discomfiture to profit from the Maxim-Schupphaus powder of which, even if patented in America and manufactured by Du Ponts, he genuinely believed he was the only true originator. His displeasure was expressed in a variety of ways. In July Hudson and Lilian were informed that their share of the rent at Thurlow Lodge was to be increased to £250 a year, 'with interest at six per cent'. This very large sum Hiram must have known they were quite unable to pay, and indeed not long afterwards the impoverished couple moved back to the more affordable accommodation offered by 47 Effra Road, Brixton.

Hiram was also, and with good reason, outraged by his brother's effrontery in claiming to have invented the aerial torpedo which he, Hiram, was on record as having patented twelve years earlier. In August, with an emphasis which must have puzzled many readers, he wrote to *The Times* from Queen's Gate Place apropos Hudson's lecture to the RUSI: 'I am receiving numerous cuttings from newspapers all over the world, some of which criticize while others give me credit for a system of aerial torpedoes of which I am supposed to be the author. However, the paper on this subject was not read by me but by Hudson Maxim, and I have nothing whatsoever to do with the subject referred to.' Relations between the brothers were now deteriorating fast, not least because Hudson was convinced that Hiram was using his influence with the Vickers company to undermine his position.

In September Hudson unburdened himself in a long letter to Jennie in Pittsfield, enclosing a copy of *The Maxim Aerial Torpedo* and complaining bitterly of the way he was being treated by Hiram and his 'powerful company':

> He and they have done their best to defeat me and if possible to drive me out of England . . . I have openly broke with Hiram . . . who was wild with jealousy and insane with envy and hatred, and did the meanest things you could possible conceive of . . . I got in a hole financially and had to sell my rights in the American powder business, at great sacrifice and loss. The Company got $175,000 for the American patents, but the whole was only to be paid in instalments of $25,000 a year. All I got out of the first instalment was $3000, and I was more than that in debt at the time . . . had it not been for Hiram's opposition, I should have at least $100,000 in my pocket.

However, he continued, looking on the bright side and ending on a more optimistic note:

> I have a [carbide] furnace now in operation in England, which is working beautifully . . . [and as to] the powder and torpedo business, I also have some good prospects of making some money . . . Hiram has lately become involved in some women matters which, if they become public, are likely to give him an unsavoury cloud over his reputation . . . Life to me has always been a race across a barren desert. I have always beheld just a little way beyond, a beautiful prospect, a rich country, watered and clothed in beautiful vegetation, but this mirage has always vanished into thin air . . . We will pray for a change.

Hudson's reference to 'women matters' reflects an inside knowledge of Hiram's private life which was soon to emerge as a weapon in the battle between the brothers. As usual where such intimacies are concerned, the precise truth is hard to come by. It had long been apparent that Sarah was not going to have children, which was a sad disappointment, for insofar as he was capable of any strong family feeling Hiram longed for a son to replace Hiram Percy in his affections. While Sarah was a devoted companion and an efficient secretary, she may have fallen short of Hiram's expectations as a sexual partner and been less than exuberant in the marital bed. However that may be, Hiram

...am (far right) and ...echanicians' during the ...nstruction of the flying ...chine in its specially-built ...gar at Baldwyn's Park, ...h work going forward on ...am engines, aerofoils and ...pellers. (Maxim Collection, ...nnecticut State Library)

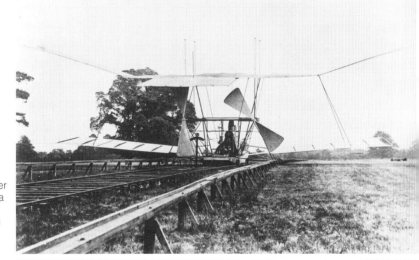

...e flying machine with outer ...roplanes' added, giving a ...od view of the track and ...surrounding trees which ...ted its length. (Vickers ...chive)

...am and his foremen Arthur ...thrie and Thomas Jackson ...vey the damage to the ...chine after the derailment ...31 July 1894. (Science ...seum/Science and Society ...ture Library)

Hiram and Sarah on the front steps of their mansion at Baldwyn's Park. Sarah not only wrote out her semi-literate husband's letters in longhand, but appears also to have signed them on his behalf. (Vickers Archive)

Hiram's son, Hiram Percy Maxim, with automobile designed and built by him for the Pope Manufacturing Company, Hartford, c1900. (Maxim Collection, Connecticut State Library)

Left: Hudson Maxim in the library of his Brooklyn home. (Hudson Maxim Papers)

Below: Hudson, Lilian and chauffeur with the inventor's latest automobile at Maxim Park, Lake Hopatcong, *c*1906. (Lake Hopatcong Historical Society)

Above: Maxim Park, Lake Hopatcong. From left to right boathouse, main residence, workshop, crenellated pier and 'fort'. (Lake Hopatcong Historical Society)

Hudson with Lilian and her father the Rev Dr Durban takes a close interest in his newly published *Science of Poetry*. Maxim Park, 1910. (Hudson Maxim Papers)

appears to have maintained his life-long habit of looking for consolation outside the family circle. Aided and abetted by Basil Zaharoff, he found no lack of outlets for his vital energies during their travels on the Continent or indeed at home, where the indications are that he did not hesitate to take advantage of the delights offered by prostitutes in the East End of London.

In the autumn of 1897 the antagonism between the brothers was exacerbated by Hiram's discovery that Hudson had for some time been negotiating behind his back with a view to selling his aerial torpedo and smokeless powder to Vickers' arch-rival, Armstrongs' Elswick Ordnance Company. Specialists in the design and construction of heavy naval guns, the firm had long been exercised by problems associated with the use of cordite, and its managing director, Sir Andrew Noble, had already indicated his willingness to conduct trials with the Maxim-Schupphaus powder in the hope that this might prove less corrosive in its effects. In September Hudson was in correspondence with Elswick about the possible manufacture by them of a 24in-calibre aerial torpedo gun to be built from his drawings, and at the same time Sir Andrew asked for 600lbs of the cannon powder to be sent over from Du Ponts for purposes of testing.

By November relations with Hiram had become so strained that Hudson decided to return to New York, taking Lilian with him and leaving her younger brother Will to supervise the management of the carbide factory. It was the first time Lilian had been abroad other than on short trips to the Continent, and she was fascinated by the hustle and bustle of the great city. While advising Du Ponts on the manufacture of his smokeless propellant at the old Brandywine powder works at Wilmington, Delaware, Hudson suggested that with their help he establish his own laboratory to carry on experiments with explosives and torpedoes. About the prospects for such a venture, however, Du Ponts were less than enthusiastic, nor despite all his blandishments did the Navy Department evince more than polite interest in his smokeless powder and other inventions. In time of peace the US authorities had neither the motivation nor the resources to give much attention to new weaponry. The navy was content to await the results of the army's smokeless powder trials, and, until decisions were taken, Hudson faced an uncertain future.

One of the few glimmers of light on the horizon was the interest being shown by the Armstrong-Whitworth company, and to encourage this Hudson took every opportunity to hammer home the superiority of his nitrocellulose-based powder over cordite, which was continuing to give cause for anxiety on grounds of safety. In March

1898 *The Engineer* carried a leading article almost certainly prompted by Hudson with the heading 'Curious troubles with slow-burning powder'. This argued that the advantages of the new propellant powders were outweighed by the risk of accidental explosions and 'the voracious way it eats out the bores of our guns . . . Nor were matters mended by the adoption of cordite, which makes it clear that if our guns have a merry life it is a short one. To such an extent has this obtained, that ships have gone to sea with a stock of ammunition which . . . would be more than sufficient to see many guns through all their lives of accurate shooting.' But the article was reassuring: given the right choices, 'prevention ought to be easily obtained . . .'.

Also in March Hudson and Lilian returned to London, disturbed by reports that in their absence Will had been unable to prevent the carbide syndicate getting into difficulties. Moving between Thurlow Lodge and Effra Road, Hudson did what he could to improve matters at the factory and to paper over the cracks in his relationship with Hiram. At the same time he met Sir Andrew Noble and Armstrongs' leading authority on naval ordnance, Josiah Vavasseur, for discussions about the Maxim-Schupphaus powder, subsequently writing to Vavasseur with regard to the patent rights. Hiram, he assured him, had said that Vickers, Sons and Maxim (which had just acquired a 40 per cent stake in the German-owned Chilworth Gunpowder Company) intended to develop their own powder. In any case he, Hudson, had made it clear that '. . . in view of the unsatisfactory treatment which I had received at the hands of his people and of himself, I did not wish to do business with them . . . under no circumstances would I ever sign another agreement with them, as every agreement they have made with me, they have broken . . .' Armstrongs were therefore free to consider his powder as an alternative to cordite, and Sir Andrew Noble pressed for delivery of the promised samples from Du Ponts.

At this critical juncture Hudson's situation, and to an extent that of the whole armaments industry, was transformed by the course of world events. As it happened the answer to the inventor's prayer came not from Europe but from another and wholly unexpected quarter, namely the island of Cuba, where a large part of the population was in revolt against the Spanish colonial government, and for years the insurgents had been carrying on a bitter guerrilla campaign aimed at causing so much confusion that the Spaniards would be glad to give up and withdraw. Spanish attempts at suppression were clumsy and harsh; there were atrocities on both sides, and the United States government tried to mediate in the face of an increasingly hostile American public stirred up by banner headlines denouncing Spanish misdeeds.

On 15 February 1898 the battleship *Maine*, despatched to the Cuban capital Havana to protect the interests of American citizens, was destroyed in the harbour by a huge explosion which killed 266 of her crew. The exact cause of the disaster has never been discovered. It is unlikely that the Spanish government was responsible, since the last thing they wanted was American intervention, but the Hearst press was in no doubt, the *New York Journal* and other newspapers being quick to allege that the ship had fallen victim to a Spanish 'torpedo mine' or 'infernal machine'.[5] As popular indignation swelled and 'Remember the *Maine*!' became the slogan of the day, a clash was inevitable. In March Congress approved appropriations of $50 million for the defence of the nation and in April Congress demanded that Spain evacuate Cuba or face the consequences. When the ultimatum was ignored, war was declared and the American army and navy prepared to take punitive action against Spanish colonial possessions in the Caribbean and the Pacific.

This, the first conflict of any consequence to involve the United States since the Civil War, could not have come at a more opportune moment for Hudson, and its effect on his fortunes was swift and dramatic. In one of its aspects, by a curious irony, the new spirit of aggression worked against his interests by holding up delivery of the samples of smokeless powder requested by Sir Andrew Noble in Newcastle. In April the American Pacific squadron under Commodore Dewey was ordered by Theodore Roosevelt to sail for the Philippine Islands, and Hudson heard from Du Ponts that in anticipation of fighting at sea merchant fleet captains were reluctant to ship any kind of explosives on which, therefore, they were imposing exorbitant freight charges. These Armstrongs refused to pay, but by this time it mattered little, for Hudson had much more important matters on his mind.

The outbreak of the Spanish-American War and the prospect of fighting on land and sea quickly served to concentrate the minds of the American military and naval authorities, and soon large orders were being placed with every available supplier of guns, shells and other war material. Directly after the sinking of the *Maine*, Hudson wrote to the Navy Department setting out the terms under which trials might be made of his various inventions relating to the propulsion of torpedoes and torpedo boats. At the same time Du Ponts were approached by the Bureau of Ordnance, which announced that it wished to conduct further trials of the Maxim-Schupphaus powder together with Hudson's

---

5. Later investigations were to suggest that her magazines had exploded following a spontaneous combustion of bunker coal.

aerial torpedo and projectiles charged with his high explosive Maximite. The company therefore renewed their agreements of 1895 and
1897 with Hudson, who was also engaged by them as 'consulting
engineer and expert in experimental work' while continuing to receive
his retainer of $500 a month.

On learning of these developments, Hiram's reaction was as explosive as one of his brother's projected aerial torpedoes. Since the British
authorities had seen fit to turn down the claims of the smokeless
powder on which he had worked so long and to the detriment of his
health, he could not expect to receive a penny for his efforts. Hudson,
on the other hand, had in his view not only succeeded in persuading
Du Ponts to adopt his powder and other pirated inventions, but thanks
to the outbreak of war now stood to make substantial profits from
their manufacture in quantity to meet the needs of the American
armed forces. As Hiram's fortunes were on the wane, so those of his
younger brother, whom he had looked down on for so long, were
suddenly set fair to change for the better. It was a bitter pill to swallow,
especially as Hudson's initiatives seemed likely to threaten the long
term interests of the newly established firm of Vickers, Sons and
Maxim.

For the war situation also carried with it political implications. According to Hiram's account it was reported in the American press that
he, Hiram, had offered to build for the US navy a small cruiser with a
large gun 'which would wipe out the whole Spanish fleet at a distance
of nine miles'. This confusion with his brother was a cause of no small
embarrassment to the Vickers board. The company owned a gun
factory in Spain, the Placencia de las Armas, and relations with the
Spanish government were cordial. Hastily a telegram was despatched
to the Spanish authorities assuring them that 'neither Hiram Maxim
nor any of his fellow-directors had anything to do with what had
appeared in the American papers, and that no such gun was in existence. This satisfied the Spaniards and our works were not interfered
with . . .' However, the company also had to insert a notice in the
newspapers to the effect that Vickers had never made any such claims,
and that 'the reports had probably emanated from someone who had
never made a gun in his life'.

As these aggravations mounted up so did Hiram's grievances, which
came to a head one day in April 1898. According to Hudson the older
man 'came to my house and told me, in the presence of my wife and
her brother [Will] that I must leave England immediately and quit
inventing anything whatever relating to naval and military matters . . .
and that I must return to my printing and publishing business, or he

would do everything he could to ruin me. He said that he was wealthy
. . . and could influence public opinion against me . . . I told him that I
was not going to go back or go down, but that I was going to continue
to go forward and go up . . . that he would learn that I would be able
to get a luncheon while he was getting a meal, and he would find ahead
of him a day of repentance and regret . . .'.

This confrontation opened an unbridgeable gulf between the
brothers, dashing any remaining hopes of reconciliation. In any case
developments on the other side of the world were such that Hudson
needed no prompting to return to America. On 1 May Commodore
Dewey's squadron of four cruisers and two gunboats reached the
Philippines and, penetrating into Manila Bay, destroyed the Spanish
fleet at anchor there in a matter of hours. Clearly the war was set to
escalate as the United States began for the first time, under the influence of the dynamic Theodore Roosevelt, to contemplate the prospect
of acquiring an overseas empire. In London Hudson was advised by
the American military attaché, General Nelson A Miles, that it would
be in his best interests to go back as soon as possible, and accordingly
Hudson and Lilian embarked on the steamship *Lucania*, taking all
their possessions with them and bidding a tearful farewell to the Durban family, who were charged with the task of noting and passing on
any sign of hostile activity on Hiram's part.

It may be that from his contacts in Du Ponts Hudson had got wind
of the cartel agreements and so realised that his chances of doing
business with Armstrongs or any other European concern were virtually nil. In any event he and Lilian had little doubt that they were
unlikely to cross the Atlantic again in the foreseeable future. Arriving
in New York, they stayed at the Waldorf Astoria, where for a few
weeks Hudson was able cut a dash on the strength of his new-found
expectations before the couple settled into a rented apartment on 34th
Street. Almost at once it became evident that Hiram's threats were not
to be taken lightly. In the matter of the disputed smokeless powder,
aerial torpedo and other devices there was little he could do to challenge Hudson's American patents in the courts. Having just lost one
legal battle the Vickers company was not prepared to risk another, and
so he vented his spleen by firing off a series of letters to newspaper
editors in England and the United States complaining that Hudson
was bamboozling the American public by posing as his more famous
brother and shamelessly taking the credit for his, Hiram's, inventions.

Hudson's first rejoinder took the form of an article in *Engineering* in
which he sought to clarify the position with regard to the Maxim-
Schupphaus multi-perforated smokeless powder. This, he explained,

was not in itself new, but the process by which its main constituent was manufactured was new, the inventors having 'discovered a new property of nitrocellulose' whereby the powder was able to achieve excellent ballistic results with safety and without having to employ a high proportion of the dangerous nitroglycerine. 'All things considered,' the army's chief of ordnance was quoted as saying, 'the [type of powder] proposed by General Rodman many years ago, and recently revived in the Maxim-Schupphaus powder, appears to me to be the most suitable and promising form for the colloidal smokeless powders.'

Under the pressures of the Spanish-American War, from April to August 1898, Hudson's approaches to the military and naval authorities at last began to yield significant results. Having again offered his high explosive and aerial torpedo to the Bureau of Ordnance, he recorded that General A R Buffington 'sent me to Sandy Hook, where . . . Maximite was subject to very thorough trial . . .' and within a short time was being supplied in artillery shells for the army. The navy also responded favourably. At the Washington Navy Yard the chief of ordnance, Captain Charles O'Neil, agreed to produce a breech mechanism for the inventor's torpedo gun, and Hudson was soon to claim that his Maximite was 'the first high explosive successfully to be fired through armour-plate, and explode behind the plate, with a delay action fuse.' Best of all was the adoption by both services of the Maxim-Schupphaus powder. Remarkably, although trials had been carried out by the army at Sandy Hook and by the navy at Indian Head, Maryland, no satisfactory smokeless propellant had yet been manufactured in quantity, and American fighting men were having to make do with the old black or prismatic brown powder.

At the decisive naval battles of Manila Bay in May and Santiago de Cuba in July American squadrons won easy victories against Spanish forces markedly inferior in gunpower, training and morale. It was, however, calculated that at relatively short ranges no more than 3 or 4 per cent of the shells fired by the US ships found their target, very few hits being obtained by the heaviest guns. This disappointing performance was put down to shortage of practice with live ammunition and also to the lack of modern guns using smokeless powder, in consequence of which the gunners' aim was constantly impeded by clouds of white smoke. During the fighting outside the harbour at Santiago only one cruiser acquired from Armstrongs, the *New Orleans*, carried smokeless powder, and the difference was clearly seen. Unusually, wrote a contemporary observer, there 'was no white cloud hanging about her . . . to interfere with the sights of her guns. This is a most important matter. We prune ourselves on the excellence of the work of our sailors, but . . . we compelled them to fight with guns that were not the best, and with powder that was as bad as any, if not the worst, in the world.'[6]

Hudson was quick to exploit his opportunity. In August, addressing the Franklin Institute in Philadelphia, he castigated American lack of military and naval preparedness, at the same time giving his version of the history of smokeless propellant:

> Probably no industry has . . . exacted more of the inventor than has the production of smokeless powder. Chemistry, physics, mechanics, mathematics and the science of gunnery, all are links in the chain upon which hangs success. Up to about 1888 there had not been produced any smokeless powder for large guns, and the importance of solving this problem enlisted in the work many of the best . . . scientific men throughout the world, many of them backed by limitless government resources or by vast private capital. Dr Schupphaus and myself entered the race . . . with the field against us. Ours was a race against the world, and we were unfortunately handicapped by . . . the difficulty of getting capitalists interested in a new and untried thing, and . . . by the dilatoriness, penuriousness and exacting methods of our own home government.

The result was that 'our troops, armed with the old, single-loading Springfield muskets [*sic*], using black gunpowder, were placed at a disadvantage in the face of the Spaniards, armed with Mauser magazine rifles, using smokeless powder. We also learned that we were very short of smokeless powder for the Navy, and that the old brown powder that we had on hand was worse than useless. The Spanish navy was destroyed mainly with cordite we got from England . . .'. The fact was that at the Navy Torpedo Station at Newport, Rhode Island, Professor Charles E Munroe had for some years been trying to adapt the new powders for heavy naval guns. Du Ponts had also been working on smokeless propellant at their research laboratory at Carney's Point, New Jersey, but their efforts were still at the experimental stage and the processes for manufacturing the powder were nowhere near completion. They therefore went on turning out prismatic brown powder under licence from the German Köln-Rottweiler company, and this it was that the services mainly depended upon throughout the war.

The purchase by Du Ponts a year earlier of the right to exploit all Hudson's patents was, therefore, timely, and he was soon pressing for

---

6. John R Spears, *The American Navy in the War with Spain* (London 1899), p256.

the adoption by the navy of 'my new explosive Maximite' and 'a cannon having a calibre of 24-inches . . . capable of penetrating the decks of light armoured cruisers . . . a veritable aerial torpedo'. The projectile for this was equipped with 'a safety delay action detonating fuse designed to explode it after having penetrated the object struck, thereby securing the maximum destructive effects.' But Hudson's first priority was to accelerate the production by Du Ponts of his 'multi-perforated cannon powder', which was soon being issued to both services. Thus in August 1898 it was stated in the supplement to the *Scientific American* that:

> An American powder, the Maxim-Schupphaus, offers a superior and scientific solution of the [military propellant] problem whose correctness has been proved by trials and tests extending over the last few years. It is the standard of the United States Army, and, after futile attempts to produce a satisfactory powder of their own, the United States Navy has lately adopted it.

Hudson's aerial torpedo was to be less successful. He was amused by, and contemptuous of, the efforts of the British and American navies to make use of the Zalinski dynamite gun which was fired, as Hiram's version of the same weapon had been, by means of compressed air.[7] This, he declared, was 'the damnedest fool contraption ever got up in the world'; nonetheless the US authorities mounted three 15in guns of this type on a light cruiser, the *Vesuvius*, which was sent to bombard Morro Castle at the entrance to Santiago harbour. 'The bombs [which were charged with 50lb of dynamite] made quite a lot of noise and spoiled a nice green grassy effect in one spot – and that was all.' This was not his only attempt to pour scorn on his brother's record as an inventor. About this time occasional items signed 'Hudson Maxim' appeared in the American press which purported to expose and ridicule certain of Hiram's inventions. When challenged, Hudson did not deny authorship, and indeed he may only have been responding in kind, for by the summer of 1898 the two men were locked into a fierce quarrel hardly equalled in its virulence since the time of Cain and Abel.

When roused each brother could be as bellicose as the other. Hiram engaged the services of a New York detective agency to keep a trace on Hudson, who for his part recalled how at this time 'a young, dapper dude of a fellow in a long coat and tall hat . . . came from London,

7. In 1892 a Zalinski gun had been mounted at Dale Fort, Milford Haven, where in trials it sank the obsolete gunboat *Harpy*.

with salary and expenses paid, to do everything he could to hurt and harry me. He visited the newspaper and magazine offices, and said he represented Hiram Maxim and his friends, and that his mission in America was to destroy Hudson Maxim'. This man 'began sending to the editors a series of really venomous circulars' before going under cover and disappearing, and Hudson was quick to fire off rejoinders. As the feathers flew, Hiram Percy in Hartford became worried about the adverse effect the publicity was having on the family name. 'You are doing yourself harm,' he wrote to Hudson in June; 'I would unquestionably ease off.' But by now the die was cast, and matters were soon to take a much more desperate turn.

By chance Hiram and Sarah had earlier in the year accepted an invitation to attend the centennial of the founding of the township of Wayne, which event took place at the beginning of August. Arriving amid a flurry of interviews and treated very much as the guest of honour, Hiram used his position to ensure that Hudson was effectively excluded from the celebrations. He then, much to Harriet's distress, became involved in a violent quarrel with Sam, who with the best of intentions had drawn up a peace agreement in an attempt to mediate between the brothers. Goaded beyond endurance by Hiram's high-handed behaviour, Hudson decided to call on all the weapons in his armoury. The brothers had always prided themselves on being men of the world, and in happier days they had felt few inhibitions about comparing notes regarding their experiences with the opposite sex. Unfortunately none of the letters they wrote to one another has come down to us, but each was well informed about the activities of the other, and whereas Hudson was unswerving in his loyalty to Lilian, it is evident that the same cannot be said of Hiram and his attitude to Sarah.

Among Hudson's personal papers there are various documents which he appears to have kept as a kind of insurance or for use as a last resort in the event of serious trouble with his brother. One scribbled sheet in Hudson's handwriting lists a series of possible points to be scored against Hiram, including the following:

- Gertie, twelve years old – critical case against Hi. Intimate with Constance also before she had her turns.
- Old Rosebud. How I met her – Hi taking her round for Zaharoff. Hi would not admit intimacy with her, but she told me he was still intimate with her.
- Hi's admission to me, in the presence of Rosebud . . . of how much more he thought of her than of Sarah . . .

The full significance of these cryptic notes can only be guessed at, but their purport is clear enough. 'Rosebud' was Victorian slang for a woman's intimate parts and was commonly used as a pseudonym by prostitutes. They are especially revealing when considered in the light of further information about Hiram's sex life provided at this time by Lilian's brother Will. Before leaving England Hudson had encouraged all his friends, including the Durban family, to look out for newspaper items and other snippets of information about Hiram which he might find helpful in the event of trouble. He did not have to wait long.

Although something of a ne'er-do-well (he had proved less than reliable as caretaker of the carbide syndicate), Will was an ardent admirer of Hudson, in whose cause he was prepared to go to considerable lengths. During the summer he wrote reporting that Hiram had seen fit to discharge a workman by the name of Hastings, who by way of getting his own back had 'threatened Hiram by letter to expose all and show the letters received by him [*ie* Hiram] from Hudson to Mrs Hi[ram] unless he received £300. Hiram said all right . . . but like a fool [Hastings] handed over the letters . . . and as soon as Hiram got them he locked them away and said to Hastings: "Out of my office. You have tried to blackmail me and if you ever communicate with me again I will get you fourteen years." ' Will's note has been kept among Hudson's papers together with what appears to be a copy of Hastings' blackmailing letter, though it is unsigned. This refers to 'the Norwood affair', and goes on to declare that: 'If I do not get the money I want sent to me . . . I shall make it my business to show you up you and your lovely Paramour also . . .'.

While certainly of interest, these unsavoury revelations were of little practical value as ammunition in the current dispute. Such matters could never be made public, nor could they ever have been adduced in a court of law. Hudson had, however, long been aware of Hiram's involvement with Helen Leighton or 'Nell Malcolm', who had returned from Colorado and had, it seems, been working as a prostitute at Poughkeepsie in upper New York. Directly after the family quarrel in Wayne, Hudson went to see Helen and talked her into laying charges against his brother for bigamy and abandonment. She needed little persuasion. Helen had never ceased to resent the cavalier way she had been treated by her former lover, nor did she fail to appreciate the financial benefits that might be expected to flow from the successful prosecution of a man who was now an international celebrity. To the consternation of Hiram and Sarah writs were therefore issued, and both sides engaged firms of New York lawyers to prepare and conduct their case.

It was not long before all the members of the family were drawn into the fray. Hiram was following a busy programme of engagements. In Boston he met Octave Chanute for the first time and discussed with him the latest developments in aeronautics, and he agreed to preside over a demonstration of his machine gun in New York. Between his appointments he wrote to Harriet asking her to attest that in matters of invention he had always taken the lead, and that Hudson had simply followed and copied his achievements. He also reminded her that, unlike Hudson, he had over the years frequently provided her and Sam with financial help. At the same time Hudson appealed to his mother and Sam to support *him* by signing an affidavit to the effect that Hiram, encouraged by Sarah, had been consistently jealous and vindictive in his dealings with his brother.

In this difficult situation Sam and the eighty-three year-old Harriet refused to take sides. Sam's son Charles replied to Hudson: 'Grandmarm . . . says that she would not sign Hiram's paper against you, and that she will not give her consent to go to such extreme lengths against Hiram . . . She does not blame Sarah for Hiram's meanness . . . but says she will not receive another cent from Hiram under any circumstances . . . he has lied so about Father [*ie* Sam] . . .' For his part, Sam had the good sense not to favour either of his embattled brothers, although he had always been closer to Hudson. As children the two of them had slept in the same bed; in their teens they had chased the local girls, given dramatised readings and interrupted church meetings to argue the case for atheism. On the other hand, and despite their differences, he yielded to no one in his admiration for Hiram, and so he excused himself on grounds of ill-health from taking any part in the proceedings.

What followed also had unpleasant repercussions for Hiram Percy Maxim, now chief engineer with the Pope Manufacturing Company, one of the leading builders of motor vehicles in the United States. In 1898 he was living with his sister Florence at the Linden hotel in Hartford, where he had been courting Josephine Hamilton, daughter of a former Congressman and governor of Maryland. In May of that year Hiram Percy proposed and was accepted, to the satisfaction of everyone including Jane, who had now married again and was no longer dependent on her clever son. During the months that followed the couple looked forward to what bade fair to be an ideal marriage, until out of a clear blue sky the young man's prospects were once again overshadowed by the malign influence exerted, albeit unwittingly, by his infuriating father.

Hiram was arrested on 7 October at the Manhattan hotel in New York and taken by detectives to Poughkeepsie, where he was released on bail of $1500 until the trial, which came on a week later. Sarah had of course always known of the threat presented by Helen. She had long since come to terms with the indiscretions associated with Hiram's private life, and she accepted without question his version of the affair, that he had been the victim of repeated attempts at black-mail of which this was simply the latest and most determined, a wicked charade masterminded by his brother with the object of discrediting him in the eyes of the American public. The press coverage given to the case was, nonetheless, a serious embarrassment. As a prominent figure in the British scientific establishment, Hiram had built up an enviable reputation which he was concerned above all to maintain. Already there were rumours of a possible knighthood in the offing, and it would be nothing short of a catastrophe if this turn of events were to put in jeopardy everything he had worked to achieve.

Not until quite late in the day did Hiram Percy become aware of the action being taken against his father, and it was with no little dismay that on 9 October he noted in his diary: 'Article came out in papers about old man's bigamy charge. Terrible shock. Makes me feel terr-ible about Josephine . . . can't express poignancy of regret.' The affair put him in a painful dilemma. While naturally inclined to favour Hudson in any dispute between the brothers, Hiram Percy was all too aware of the disagreeable consequences for the family and particularly his mother if Hiram were to be convicted. Jane was so upset by the shaming implications of the charges brought against her former hus-band and by the humiliation of being called to give evidence that Hiram Percy feared for her health and did everything he could to shield her from the unavoidable publicity.

He also feared for the future of his engagement until speedily reas-sured by his fiancée, recording next day: 'Received letter from Josephine saying she would stick by me, but I can't get over the awful suffering it must be causing her.' Before and during the trial Hiram Percy was approached by both sides to give evidence, but this he firmly declined to do. Subsequent diary entries[8] tell their own story:

> *13 October*   Special delivery from Hudson asking me to be arbiter.
> *16 October*   Nothing in morning papers, thank God! Thinking of some way to head off any papers mention of Mother. Received copies of papers, probably from Sarah . . . I ought to change my name, it seems to me. Wrote to Sarah Haynes [*sic*] about stopping further paper scandal . . .

---

8. Quoted in *Family Reunion*.

17 October Talked over situation [with Hiram's attorney]. Old man very evidently in last stages of despondency. Wrote letter to me supplicating. Wonder if I am to come out all right! Cannot think that I am to lose all so undeservedly. Have confidence in our coming out clean, but feel mean over it.

20 October [After conferring with lawyers he was still] worrying about outcome of hearing . . . but found impressions were right about nothing coming from case.

This was indeed so. On 15 October, under the headline 'Maxim no Bigamist', the *New York Herald* reported that the charges brought against Hiram had been dropped. Helen 'became confused' when cross-examined about the details of the wedding ceremony alleged to have taken place in 1878, while Hiram's counsel drew from her the admission that 'she had lived in several disorderly house . . . and at one of them went by the nickname of "Tug Wilson"'. It was pointed out that Hiram had for many years been divorced from his first wife, Jane, who had meanwhile remarried, as had Hiram, whose second wife, 'a tall, handsomely attired woman', was present in court. The charges brought by Helen were therefore found to be groundless and the case was dismissed, as was a claim for damages. At the same time Hiram's assertion that Helen had been induced to bring the action by Hudson was discounted by the court, for 'even if it were shown that the brother had urged her on, it would not add to or detract from the merits of the case'.

For Hiram the verdict came, naturally, as an enormous relief, appearing as it did to justify his claim to have been the victim of a sordid attempt at blackmail. The subsequent fate of the unfortunate Helen is not recorded. Earlier in the year Romaine, now nineteen, had expressed a wish to see her natural mother, but in the circumstances this was not deemed advisable. As for Hiram Percy and Josephine, they were married in December and spent their honeymoon in Europe. Needless to say they avoided calling on Hiram and Sarah, although when in London their curiosity did get the better of them and they are said to have 'peeped at' Hiram from an adjacent building as he went into his office. Their union was to be a conspicuously happy and successful one, ending only with their deaths within a few months of one another in 1936.

The feud between the brothers rumbled on, having its effect not only on the Maxim family but also on aspects of the wider scene, notably the attitude of the British and American naval authorities towards the smokeless propellants used in service projectiles. In private conversation Hiram continued to display a hostility towards Hudson and, by association, Hiram Percy which amounted to paranoia. In his public utterances he was more cautious, confining himself to making veiled references to a 'double' in the United States who was 'masquerading as Maxim, the gun inventor'. In *My Life* Hudson is nowhere mentioned by name, although its author describes with heavy emphasis how in 1888 his request for workmen to be brought over from America was 'the cause of one of the greatest misfortunes of my life; in fact . . . the greatest trouble that I ever had, as it brought into my life an individual who caused me an immense amount of vexation . . . and the loss of many thousands of pounds.' Again: 'I often see in the newspapers that I am lecturing on high explosives in the States, that I am bragging about my guns, that I am burning nitro-glycerine in a lamp or doing some other foolish trick, when I know all the time that I am in Europe. It may be laughably funny to have a double – but I find it a decided nuisance.'

After 1898 Hiram and Hudson cut off all contact with one another, pursuing their separate careers on opposite sides of the Atlantic. Hiram and Sarah were to pay only one more short and troubled visit to the United States, while Hudson and Lilian were not to return to England until more than twenty-five years had elapsed. Harriet and Sam, having given offence by declining to support either of the brothers during the bigamy trial, were virtually ignored by both of them until a combination of the imminence of death, the shame of poverty and the consciences of their wives obliged them to relent. Hiram Percy continued to be on friendly terms with his uncle Hudson, to whom he grew closer over the years, but he was always too busy to make the journey to see Harriet and Sam at Wayne. His father he was never to see again.

SIX

Fame and Fortune

Property in patents is not respected by the majority of mankind as other property is.'

HIRAM MAXIM

Even though he had been acquitted of the charges brought against him, Hiram had reason to be thankful that events taking place in the United States found little resonance in the old country, and that editorial discretion could be relied upon to ensure that no reference to the bigamy case appeared in the British press. Dr Durban was therefore unable to dig up much in the way of gossip sufficiently interesting to pass on to his son-in-law, having to content himself with general impressions as, for example, of Hiram's reputation in City circles: 'Morally,' he wrote to Hudson in March 1899, 'his name is loathed there. In fact nobody would condescend to mention him were it not for his scientific genius.'

The controversy over smokeless powder patents simmered on, with the main protagonists scoring points against one another in technical journals on both sides of the Atlantic. In May 1899 Hudson penned a long letter to the *Scientific American* describing in detail the genesis of the Maxim-Schupphaus powder, whereupon Hiram wrote indignantly denying his brother's assertion that he had contributed to his early work on propellants: 'I am glad,' retorted Hudson, 'because this refutes his previous claims that the inventions which I have patented were on ideas acquired by me from him while assisting him with his experiments.' Nor did Hudson scruple to suggest that Hiram's researches owed much to his rivals' work on the cordite patent, notwithstanding the fact that he had 'rushed into the Patent Office ahead of Messrs Abel and Dewar, and lodged his provisional application about fourteen days ahead of theirs'.

All this was trying enough, but as Hiram approached his sixtieth year he had other reasons for dissatisfaction. His quest for a flying machine had, like so many of his endeavours, involved a great deal of time and effort but yielded no practical or commercial results. Several of his allies in Vickers, Sons and Maxim, including Symon and Cloete, had retired or moved on, and, although he continued regularly to attend board meetings, the measures taken by Sigmund Loewe had drastically reduced his influence within the company. There were, however, consolations. Following the worldwide adoption of the Maxim gun he was internationally famous and, despite the large sums he had spent on his aviation experiments, by any standard a wealthy man. After 1898, moreover, both his bank book and his reputation were to be enhanced, as Hudson's had been, in consequence of exciting developments taking place in the remoter parts of the world.

In 1894 the School of Musketry at Hythe was still describing the Maxim gun as 'a somewhat puzzling addition to the machinery of the battle line in civilised warfare', but two years later Captain C E Callwell RA of the Intelligence Division of the War Office published a monograph entitled *Small Wars, their Principles and Practice*, in which he devoted two pages to the machine gun. Until recently, he declared, these had been clumsy and unreliable, while their tactical role was 'not yet finally established'. With the introduction of the world standard Maxim gun the situation had changed, for it had demonstrated its value in a number of colonial campaigns and there was 'every reason to believe that this class of weapon has a bright future before it'. Particularly when used against primitive opponents, Maxims were 'undoubtedly most efficaceous . . . the effect of such weapons against rushes of Zulus, Ghazis or other fanatics [being] tremendous if their fire is well maintained'.

By March 1898, when a feature article appeared in *Engineering* underlining the power and versatility of the Maxim gun, it was evident that the message was getting home. Already General Kitchener was well advanced in his preparations for the re-conquest of the Sudan, his expeditionary force of 26,000 men being supported by 80 field guns and 44 Maxims. Advancing by stages up the Nile, the combined British and Egyptian army finally, on 2 September, confronted the enemy outside the city of Omdurman, where two batteries of Maxim guns took a frontal position against an estimated 40,000 tribesmen. Led by the Khalifa in person and making up in courage what they lacked in firepower, the warriors rushed forward to the attack, confident of divine protection. Of the battle or, more accurately, massacre that followed, the newspapers reported that as the machine guns opened fire 'a visible wave of death swept over the advancing host', and it was estimated that three-quarters of the 20,000 dervishes who fell were mown down by the Maxims before they reached the British lines. At a cost of five officers and 85 men killed the death of General Gordon at Khartoum had been avenged.

'In the past,' famously commented the correspondent of *The Times*, Sir Edward Arnold, 'our wars have been won by the dash, the skill and the courage of our officers and men, but our last campaign has been won by a very quiet, scientific gentleman living down in Kent.' This description of Hiram doubtless caused some hilarity among friends and colleagues who were prepared to put up with his overweening conceit only because he was a genius and an American and therefore could not be expected to behave like normal people. Still, the part played by the Maxim gun in so resounding a victory was widely acknowledged in the cartoons and popular literature of the day, and not long afterwards the idea was mooted of honouring the inventor for his services to the world of science and to the Empire. In due course a discreet approach was made to Hiram with the offer of a knighthood, it being pointed out to him that since he was an American subject this would have to be of an honorary nature only.

While naturally gratified, Hiram was not entirely happy with this proposal. Over the years he had been irritated as much by the indifference shown by the American authorities towards his inventions as by his dispute with the Colt company and Hudson's success in exploiting his smokeless powder and other patents. By contrast the British establishment had from the outset recognised his achievements and enabled him to profit from them, and since he now felt more at home in England than in the United States he decided to become naturalised in order to qualify for the full knighthood. This step he justified on general as well as pragmatic grounds. 'The reception that I received in England,' he wrote in *My Life*, 'and the straightforward honesty of the gentlemen with whom I had to deal, gave me a very favourable opinion of the British character, and it occurred to me, especially after I had met the then Prince of Wales and other members of the Royal Family . . . that I ought to become a British subject, especially so as I was permanently settled in England.'

The naturalisation process having run its course, the award was announced in the New Year's Honours list for 1901, the first of the new century, and Hiram looked forward to receiving the ultimate accolade by the hand of Queen Victoria. In the event the venerated monarch died a few weeks later, and so it was that on 9 February 1901 the London Gazette announced that: 'The King was this day pleased to confer the honour of Knighthood on Hiram Stevens Maxim Esq.' Messages of congratulation poured in from all sides, not least from the many learned societies to which Hiram belonged, and the newspapers pronounced him the only new knight of any real interest, the others being a dull lot. Once again the British establishment had demonstrated its readiness to break from convention and recognise talent from whatever source it came. It was also the climax of an extraordinary career, marking a storybook transition from rags to riches. The investiture was one of Edward VII's first public duties, and his pleasure was genuine enough. The new king had admired Hiram's style ever since the two men had first met fourteen years before at Hatton Garden, and by this time the Maxim gun had again demonstrated its worth on the field of battle.

During the early stages of the South African War, which broke out in October 1899, the Boer republics were able to put into the field more men and better weaponry than the British forces opposed to them. For some time President Kruger of the Transvaal had been importing Mauser rifles and Krupp and Creusot field guns, and with the acquiescence of the War Office the Boers had also purchased through their agents in Europe numbers of the 37mm Maxim gun soon to be dubbed by the African natives the 'pom-pom'. This the British troops did not possess, because to Hiram's annoyance the military adviser to Maxim Nordenfelt, Captain Acland (whom he referred to as 'Captain Calamity') had chosen to advise the War Office against adopting the heavier weapon for field service. At first the hard-pressed British could do little more than hold their ground, and when in December they tried to go over to the offensive they sustained heavy defeats at Stormberg, Magersfontein and Colenso. With the government facing a public outcry, the commander on the spot, Sir Redvers Buller, telegraphed for more men, more artillery and more Maxim rifle calibre machine guns. Belatedly, he also called for the 'pom-pom', which he described as a 'wonderful weapon' in the hands of the Boers.

As a result Vickers, Sons and Maxim were provided with a welcome opportunity to expand their capacity and to compete with Armstrong-Whitworth and the Royal Arsenal at Woolwich as a leading supplier of munitions to the army as well as to the Royal Navy. Between October 1899 and May 1902 the company supplied nearly £2½ million worth of military equipment to the War Office, and in the first two years of the war fifty 37mm 'pom-poms' and several hundred rifle calibre Maxims were sent out to South Africa. Unfortunately there was a downside in that to meet this unexpected demand many foreign orders which would normally have been met by the Erith and Crayford works had to be passed on to the German firm of Deutsche Waffen. Although this led to the loss of some foreign customers and had the effect of cutting back Vickers' profits, there was no question but that priority had to be given to the needs of the army.

Regular courses of instruction in the Maxim guns having been introduced at the musketry training schools, these weapons made their contribution to a rising volume of firepower which overwhelmed the Boer commandos, obliging them to disperse and resort to guerrilla warfare. The ultimate British victory was, however, mainly due to the deployment of a better and more numerous artillery, including 15-pounder Ehrhardt quick-firing guns purchased at short notice from Germany. The reaction of senior officers to the performance of the Maxim and other machine guns such as the Colt and Hotchkiss was mixed. The 'pom-pom', regarded as an adjunct to the artillery, was valued by some on account of its explosive shell, which could be seen bursting and had a demoralising effect on the enemy. Others thought that while valuable in defence it was no substitute for field guns, the shell being too small to do much damage, and one gunnery officer spoke for many when he described it as 'an invention of the devil . . . I can't find either its place or its utility except as a range-finder. The enemy had it, so we had to get it too to give confidence, but [its effect] is almost entirely moral.'

As to the rifle calibre machine gun, this was thought valuable mainly as a defensive, infantry support weapon. Under the conditions of the African veldt, with a swiftly-moving enemy taking advantage of every available scrap of cover, targets were not always easy to find and the tendency was to shoot off large amounts of ammunition with little apparent result. Nevertheless, there was no doubting that the Maxims were good at keeping the enemy's head down and for repelling an attack in the open, and one officer declared that 'to fight any body of infantry unsupported by machine guns is nothing short of a crime'. The consensus was that an allocation of two Maxims to each infantry battalion was about right, and after the war this was confirmed by the War Office as at least a nominal objective.

In this the British army was one step ahead of all its rivals with the exception of the Germans and Austrians, whose war departments had been among the first to place orders for Maxim guns. Subsequently, as already noted, Krupps of Essen and Deutsche Waffen of Berlin had acquired the rights to produce Maxims under licence for the German army and navy, but few guns were actually delivered until Major von Tiedemann, the German military attaché, was able to see for himself the havoc caused by the weapon at the battle of Omdurman. Within a year Maxims were being issued in numbers to the Imperial army, and Vickers started to earn substantial royalties on the manufacture of guns in Germany. The accounts of the company's Maxim Branch for 1899 include payments of £27,000 from Krupps and £18,000 from

Deutsche Waffen, although its overall net profit amounted to only £31,000. In that year Hiram earned £3500 in salary and bonuses (less than half the sum paid to Sigmund Loewe and Basil Zaharoff) and all the directors benefited proportionately when in 1900 Deutsche Waffen also undertook to make Maxim guns for the Russian government, whereupon the Branch's net profit rose to £172,000 before falling back to £147,000 in 1901 and £88,000 in 1902.

By the time he received his knighthood, therefore, Hiram's income deriving from sales of the Maxim gun was such as to ensure financial security for the rest of his life. Back in New England, and especially in Maine, the news of his elevation to the ranks of the English aristocracy caused something of a sensation, and at the family homestead in Wayne Harriet, now eighty-six, expressed a wish to be reconciled with her famous son and to see him once more before she died. Accordingly in the spring of 1901, responding to a letter from Sam, Hiram and Sarah again made the Atlantic crossing. Sadly, it does not seem to have been a happy occasion. On his arrival in New York Hiram was approached by a group of businessmen who announced that they proposed to form an American Maxim Gun Company, and asked him to sign a disclaimer to the effect that he would not stand in the way of the enterprise. Suspecting that his brother was involved, Hiram refused, pointing out that he had no right to approve a subsidiary company without the consent of his fellow directors, and so the proposal came to nothing, although it was to be revived many years later with unfortunate results.

It is characteristic of Hiram that in his autobiography he was at pains to denounce his brother's supposed infamy in the matter of the American gun company, but that he made no reference to the meeting with his mother which was the reason for his visiting the States. He and Sarah travelled to Maine to see Harriet and Sam, but the family quarrel was too fresh in all their minds for their stay to be free of awkwardness. Sir Hiram, pursued by reporters, indignantly repudiated rumours put about by a 'contemptible blackmailer' to the effect that during the Civil War he had deserted from the army and run away to Canada.[1] For these rumours Hudson was indeed responsible, but Hiram was careful to guard his tongue for fear of becoming involved in fresh legal entanglements, and after a short time he and Sarah were glad to return home.

A few weeks later, though she was according to Hudson 'still light on her feet, quick-witted, and full of the joy of living', Harriet

---

1. He obtained an affidavit from the authorities in Maine confirming that since he had never enlisted in the US army he could not have deserted from it.

succumbed to an attack of pneumonia and was buried beside her husband in the churchyard at Wayne. Neither Hiram nor Hudson appears to have been present at the funeral, and it was left to a local historian to compose an appropriately ringing epitaph:

> None but one possessed of great physical structure could have taken her place and achieved so great a victory. Napoleon once said, 'Great men have great mothers'. Whose mother is justly entitled to greater honor, for who had produced greater men than she?

Back in England Sir Hiram and Lady Maxim settled down to enjoy the benefits of their new social status, which was soon presenting them with scope for wider responsibilities. In the autumn of 1901 it was suggested to Hiram that he stand as a Conservative parliamentary candidate in the general election due to take place the following summer. This he had the good sense to decline, mainly on account of his deafness but also because, as he well knew, he would have felt out of place in the House of Commons. Instead he chose at least for the moment to opt for a relatively quiet life, the better to enjoy his share of Vickers' profits from the conflict in South Africa, the more welcome since they had been so long in coming.

For others the war was proving both inconvenient and costly. As the guerrilla fighting dragged on into 1902 business interests were suffering from the closure of the Rand goldfields, and Hiram was approached by friends in the City (possibly because of his neutral American background) who asked him to intercede with a delegation of Boer leaders visiting the Hague. Authorised to offer generous financial inducements if the Boers would lay down their arms and allow the gold mines to resume production, Sir Hiram and Lady Maxim went over to the Dutch capital and did their best, pointing out that, since the British were bound to win, their foes would be well advised to take the money and avoid further pointless suffering. Having listened politely, the Boer officials replied that they had every faith that God would support their cause, and so nothing more could be done. A few weeks later the Peace of Vereeniging was signed and the war came to an end.

Hudson, meanwhile, was taking full advantage of the determination of the United States government to enlarge and modernise its armed services in the light of the shortcomings exposed by the Spanish-American War. As in other maritime nations this endeavour was led by a programme of warship and particularly battleship construction, expenditure being driven upwards by the influential Navy League composed, in the words of a leading naval historian, 'of naval officers, disinterested patriots, and representatives of firms that might be expected to benefit by an increase in armament expenditures'.[2] Before 1898 the annual sum allocated to the navy had seldom exceeded $30 million, but, with the coming of hostilities and the subsequent commitment of the United States to consolidating its position as an imperial power, appropriations increased rapidly, rising to $77 million in 1900, $115 million in 1905 and $147 million in 1910.

An important consideration was that war material be produced in the United States rather than imported, as so often in the past, from Europe. Reliance on local sources offered obvious advantages in terms of security while encouraging domestic manufactures and avoiding the payment of royalties to foreigners. There were also strategic and political implications. As the Americans became more ambitious in terms of maritime trade and imperial expansion, so the risk of war with Japan and/or a combination of European powers had to be reckoned with as a possibility. In these circumstances it was clearly unwise to be dependent on overseas suppliers and the authorities, particularly the Navy Department, preferred to deal with American companies such as Du Ponts and home-grown consultants such as Hudson Maxim. The latter offered access to significant areas of expertise, and the fact that many of his ideas probably derived from his celebrated brother was not seen as a problem.

After 1898 strenuous efforts were made to raise the standard of equipment supplied to the American services to the level of their European counterparts, and Hudson was one of a number of specialists consulted by the Board of Ordnance with a view to improving the efficiency of guns, explosives and projectiles. At first he met with disappointment when the aerial torpedo on which he had worked for so long was rejected by the Navy Department as impractical, tests having shown that, even when carrying a large explosive charge, such a missile had little destructive effect on the hull of an armoured ship. On the other hand both the Maxim-Schupphaus smokeless powder and the inventor's Maximite, a picric acid-based high explosive[3] similar to the lyddite and melinite widely used in Europe, were soon to be accepted as the standard propellant and bursting charge in the shells of the army and navy.

At the Torpedo Station at Newport, Rhode Island, the navy's own specialists, led by Professor Munroe and Lieutenant John B Bernadou,

---

2. Donald W Mitchell, *History of the Modern American Navy* (London 1946), p136.
3. It was in fact composed of picric acid and mono-nitro-naphthaline.

had also been working on propellants, coming up with 'a pure colloid produced by the action of ether and alcohol on soluble gun cotton'. Partly because neither this nor Hudson's powder (as later modified) contained the highly suspect nitroglycerine, samples of both were ordered by the army and the navy. A new government factory was built at Indian Head, Maryland, to produce the naval propellant, but in the event by far the lion's share of orders went to Du Ponts for the manufacture of the Maxim-Schupphaus powder. As the Secretary of the Navy observed in his Report for 1899:

> The Government powder factory at Indian Head is progressing favourably . . . [though] it is neither expected nor desired to enter into competition with private manufacturers . . . it being the policy of the Department to foster the commercial industry, upon which the country must largely draw its supply.

Following successful trials the Bureau of Ordnance and the Navy Department confirmed their adoption of the Maxim-Schupphaus powder, which was thought to be as effective in heavy guns as the Anglo-German propellants while being safer and less liable to erode the gun barrels. Not only did it give higher muzzle velocities than the old gunpowders, with greater range and accuracy, it enabled more shells to be carried in the magazines, and so the powder was introduced as standard for all naval ammunition including projectiles for the 12in guns of the new battleships *Maine*, *Missouri* and *Ohio*. The efficiency of these weapons, commented the *Scientific American* in December 1900, 'speaks volumes for the excellence of the multi-perforated, all-guncotton smokeless powder which has been adopted by the navy; for unlike the high nitro-glycerine powders such as cordite . . . our new navy powder achieves these splendid results [while minimising the problem of erosion]'.

Nor was the navy slow to follow the example of the army and carry out tests of the high explosive Maximite, some of which was loaded into shells armed with a delayed-action fuse devised by Hudson and Captain Dunn of Frankford Arsenal. In these as in other of his activities Hudson did not hesitate to draw on his brother's earlier patents, but, although Hiram was furious when he learned of what was going on, there was little he could do. Many scientists on both sides of the Atlantic were trying to come up with a reliable fuse for armour-piercing projectiles, while, as for Maximite, the production of picric acid-based high explosives was well established: 'This new and powerful explosive,' wrote Hiram, 'which was first discovered in France and

called "Melinite" . . . was re-invented by many others later on and
called "Dunnite", "Smithite", "Jonesite" or "Bugginsite", always
bearing the name of the last man who re-invented it. It was,' he added,
'so simple a matter to prepare this explosive that anyone could do it.'

During the course of 1901 heavy naval shells filled with Maximite
were tested at the New York Arsenal in competition with a large
projectile similar to Hudson's aerial torpedo but invented by one
Louis Gathmann of Chicago, who claimed that his version was capable
of sinking opposing vessels by shattering their protective armour.
Hudson, an interested onlooker, noted with evident satisfaction that:

> The eighteen-inch Gathmann shell weighed nearly a ton and carried
> about five hundred pounds of guncotton. It had a soft nose which
> was to collapse on the plate and make the shell explode. At the first
> shot the projectile struck the [target] plate squarely and exploded,
> but the only effect on the plate was to leave a great yellow smudge on
> its face . . . Then a twelve-inch armour piercing shell weighing two
> hundred and fifty pounds and carrying only twenty-three pounds of
> Maximite, and equipped with a delay action fuse, was fired at the
> other plate. It exploded in the plate when about two-thirds through,
> blew a hole as big as a barrel, and shattered the plate into fragments.

Hudson's account is supported by the official report of the Bureau
of Ordnance, and also by Professor Alger, who commented that to the
surprise of those present the effect of the Gathmann shell on the
target was 'trifling'. 'We can safely conclude,' he went on, 'that [such a
shell], even if of the largest practicable capacity, exploding against the
armoured side of a vessel would be practically harmless.' The
Gathmann shell was, therefore, rejected and Maximite was approved
by the navy for use in armour-piercing high explosive projectiles,
supplementing traditional black powder-filled ammunition. As the
New England Magazine confirmed, the new explosive had been taken
up by the authorities on account of its stability, its powers of fragmen-
tation and its ability to pass through armour plate before exploding.

Very soon Hudson was riding on the crest of an irresistible wave of
good fortune, which went far to compensate for his chagrin on hearing
of the award of Hiram's knighthood. Even as his smokeless powder
went into full production, the *Army and Navy Journal* was noting (in
November 1901) that ordnance officers had 'formed a very favourable
opinion' of Maximite, whose inventor 'appears to have obtained
remarkable success in uniting insensitiveness with . . . immense explosive force per unit of volume. Shells loaded with it have been fired

through armor plates from three to twelve inches thick . . . showing that it will stand the shock of going through any armor plate that shells loaded with it will penetrate.' Du Ponts looked forward to a steadily rising volume of orders for smokeless powder and Maximite, over which they had a near-monopoly of manufacture, and Hudson's commission on the profits rose in proportion.

Exact figures are hard to come by, but in January 1902 it was reported that the army had purchased 75,000lbs of Maximite, and there seems no reason to doubt Hudson's later claim that initially the US government 'bought of me [that is, Du Ponts] the secret of Maximite for fifty thousand dollars, and soon afterward the Navy Department bought an amount . . . which gave me a profit of eight thousand dollars.' During the coming years the inventor was to modify and refine the composition of his smokeless powder and high explosives, from which he continued to earn a healthy income from royalty payments. At the same time, however, he was busy with yet another project which had long been engaging the attention of the world's navies.

In addition to battleships and cruisers the Department of the Navy had commissioned twelve torpedo boats, and it was, therefore, anxious to encourage the production of a serviceable torpedo of American manufacture. During the Spanish-American War the ineffectiveness of the torpedoes launched by both sides had been amply demonstrated. The Whitehead automobile or 'fish' torpedo in general use during the 1890s still had a range of little more than 1000 yards, and on the few occasions it succeeded in hitting the target its small charge of guncotton could inflict only minor damage. Like other inventors, including his brother, Hudson had long been fascinated by what he saw as the neglected potential of this weapon, and in 1898 he approached the Navy Department with the formula for an improved propellant which, he suggested, could be used not only in the standard Whitehead torpedo but also for driving motor torpedo boats and other small vessels.

Although the manufacturing rights to the Whitehead torpedo had earlier been acquired by the E W Bliss Company of Brooklyn, makers of canning machinery, the firm had been given little encouragement to exploit them. With the outbreak of war, however, Hudson consulted the firm's chief engineer, Frank M Leavitt, with whom he was soon collaborating on a new variant of the weapon incorporating a turbine engine as recently developed in England, and, as an alternative to the compressed-air motor of the Whitehead, a heater motor driven by steam generated under pressure by a combustible fuel. While Leavitt was interested mainly in the turbine engine, Hudson concentrated on the propellant fuel, and in 1899 he filed a patent for an apparatus for producing a motor fluid 'mainly for the more rapid propulsion of self-propelled torpedoes, torpedo boats and light naval vessels'. Others soon followed for 'a means of launching torpedoes so that the boat can launch them as near as possible to its target, i.e. from a tube . . . already immersed in water', and for 'saving space by using a heater motor, so enabling the torpedo to run faster and carry a larger explosive charge'.

It was the beginning of an eventful working relationship with the Leavitt and Bliss company, the success of which depended on developing a motor and fuel capable of propelling torpedoes to a more effective range. To this challenge Hudson rose with all his brother's dedication, spurred on by the decision of Congress to order six Holland-type submarines for the navy. Early in 1901 he registered his patent for a 'motive-power combustible for automobile torpedoes, designated "motorite"', and the response from the navy and from Du Ponts was encouraging. In April Eugene du Pont informed the inventor that the company was prepared to provide him with research facilities at its Forcite Powder Works at Lake Hopatcong, New Jersey, on condition that 'if the Motorite is a success, we will have the manufacture of same.' Situated some fifty miles west of New York, this was remote enough for experimenting with explosives, had good access by rail and was close by the army's main powder arsenal at Picatinny.

On his first visit to Lake Hopatcong Hudson could not help being struck by the beauty of the place. The lake was surrounded by well-wooded, low hills which sloped gently down to the water and were little inhabited apart from a few scattered dwellings and the occasional wayfarer or hunter after wildlife. The powder works, however, were the real attraction, and he wrote that 'if they'd been in a desert, that wouldn't have deterred me. I came primarily that I might have room to grow up and blow up with the country without too much interference with neighbours.' At once he rented a cottage for himself and Lilian on the shore of a sequestered bay, engaged a local man, William Sperry, as his assistant and set about building a laboratory and workshop. When his experiments were sufficiently advanced, an agreement was negotiated whereby the Bliss company would supply the torpedoes and Du Ponts would manufacture the propellant fuel. It was all something of a gamble, for there was no guarantee that the finished product would be acceptable to the Navy Department, and in the meantime Hudson had to subsidise the work from his own pocket. He later claimed to have spent $75,000 on his torpedo and motorite researches, which outlay was covered by his income from Du Ponts and the sale to the Union Carbide Company of his process for making calcium carbide.

During the next twelve months, and using the lake as a test site, the inventor and his team of assistants conducted experiments with propulsive fuels and explosive warheads. That it was dangerous as well as difficult work was underlined in October 1902 when their workshop was wrecked by an explosion which killed William Sperry. This was, however, only a temporary setback. It was reported that the local coroner, having viewed the scene of the accident, considered an inquest unnecessary, and soon afterwards Hudson announced to Pierre S du Pont that he believed he had a torpedo which met the specifications laid down by the Naval Torpedo Station. As it happens the brothers Pierre, Alfred and Coleman du Pont had earlier in the year taken over the running of the company, and under their energetic leadership plans to reorganise the US explosives industry were already showing results. Within a very short time Du Pont was to consolidate its position as the principal private manufacturer of smokeless powder and other military explosives, and in consequence the firm was taking a growing interest in the development of weaponry for the armed services.

Between 1902 and 1904 trials of the Whitehead torpedo as modified by Leavitt and Hudson and driven by compressed air or by 'motorite' were conducted by the Naval Torpedo Station in competition with other contenders. Unfortunately for Hudson, his propulsive system was open to the objections that motorite contained a high proportion of nitroglycerine, which was deemed unsafe, and that it was no more efficient than the compressed air motor it sought to displace. The Navy Department therefore came down in favour of Leavitt's design which retained the compressed air motor. This, designated as the Bliss-Leavitt Mark I, was adopted in November 1905, the first of a series of such weapons which were to remain in the American service up to and during the First World War. The Department did, however, agree to purchase for a token sum the secret of motorite, partly to retain an option on its manufacture, and partly to prevent anyone else from acquiring it. Hudson was always to maintain that the navy 'bought his invention for a motorite-driven torpedo', but in the event, despite repeated promptings by him, no such torpedo was ever introduced the service.

Despite this and other disappointments, sales of Hudson's smokeless powders and high explosives were such that the Maxim lifestyle was soon to be transformed. In 1902 he and Lilian bought their first house at Sterling Place in Brooklyn, 'a very dear little place'. It cost

$8000 and had two stories and a basement, but its main appeal for Hudson consisted in the 'thick-walled brick laboratory with no windows and doors' which he built in the backyard. Here he was able to carry on his researches with the assistance of Lilian, who, 'in spite of a natural timidity . . . never hesitated to help me in dangerous experiments'. She does appear to have been in all respects the perfect wife, and Hudson responded to her with an affectionate ardour which illumines the long pencilled scrawls he sent to her every day they happened to be apart. About the intensity of their relationship there can be no doubt, although there were no children, and whether or not this was the result of a deliberate decision remains unknown.

In 1903 Hudson acquired his first automobile, which was notorious for breaking down but did enable him to commute more easily between New York and Hopatcong. It was the first of a succession of ever more powerful and expensive vehicles which to the dismay of his passengers (especially Lilian, who insisted that he hire a chauffeur) he took pride in driving at breakneck speed over the dusty, pot-holed roads which then made life difficult for motor and motorist alike. Delighted by Lake Hopatcong and its environs, he and Lilian decided at an early stage to invest in a parcel of land by the banks of the strangely-named River Styx, where they proceeded to build a house for themselves and to develop an interest in real estate. Before long they had formed the habit of spending the summer months at Hopatcong and the winters in New York, a pattern they were to follow for most of Hudson's working life.

In 1903 Robert Schupphaus announced his intention of returning to Germany, whereupon Hudson bought his interest in the Maxim-Schupphaus powder for $1000 down and a further $10,000 to be paid over three years. It was a sound move. By 1904 Du Ponts had established a near-monopoly of military smokeless propellant, only a small amount being made by government-owned and operated works, and Hudson benefited accordingly. In 1905 the company also agreed to pay him $115,000 for the sole rights to produce his smokeless powder in the United States, the inventor to receive 20 per cent of net profits on sales. A similar deal having been done with the Mexican government, Hudson went to Santa Fe to advise the Mexican authorities on its manufacture, and in September he assigned to Du Ponts all his foreign patents for $57,500. These payments were made over a period, but added to commissions on sales and his retainer of $500 a month they amounted to a substantial sum. Hudson was later to claim that during the first ten years of the century he earned on average over $50,000 a year, and this is probably no exaggeration.

Not all the couple's income derived from Hudson's inventions. Reasoning that as the public grew more accustomed to travelling by road as well as rail there would be a rising demand for holiday homes, they used part of his earnings from Du Ponts to buy more land around Lake Hopatcong. This they divided into plots which were sold on long leases to well-to-do New Yorkers looking for a weekend or summer base for fishing and boating or for retirement. From the outset Lilian took a large share of the administrative responsibility this entailed, her good looks and charm helping to find customers for the plots and to persuade the leaseholders into accepting the high standard of building and maintenance necessary to make each property as attractive as possible, and enhance the value of the area as a whole. By 1905 the enterprise, known locally as Maxim Park, was showing a sufficient return for the owners to plan further extensions to their own already imposing dwelling, to erect a crenellated stone pier leading to the boathouse and to acquire a hotel, the Hotel Durban, for accommodating shorter-term visitors.

In 1906 Hudson and Lilian moved in New York from Sterling Place to a larger house on St Marks Avenue, Brooklyn, which, decorated with Maxim guns, shells and fuses, was conveniently situated within walking distance of the Navy Yard. Here over the next few years the inventor and his attractive wife built up a circle of friends influential in the arts and sciences, and Hudson contrived to become something of a minor celebrity. Conscious, like his brother, of the value of being in the news, he had long been an assiduous contributor to popular as well as technical journals, putting forward ideas which were often perceptive and on occasion surprisingly original. Thus as early as May 1889 the *Scientific American Supplement* had carried an article by him entitled 'The principle of force and demonstration of the existence of the atom', which he was to maintain had anticipated the general atomic theory propounded by later generations of physicists. Far-fetched though this might seem, Hudson's claim is given credence by no less an authority than the *Encyclopaedia Britannica*, which confirms that he 'formulated an hypothesis on the compound nature of atoms not unlike the atomic theory which was later to be generally accepted.'

Apart from his status as a leading explosives expert, Hudson was now reading widely, building up extensive libraries in both his homes and aspiring to recognition as a kind of all-round authority on scientific, artistic and intellectual matters. Rarely did he pass up an opportunity to express himself in public, and he soon became known for his trenchant opinions on a variety of topics such as pacifism (which not surprisingly he was against), rearmament (which he was for) and womens' rights (also for). As today he would have appeared regularly on television and radio chat shows, so then he was in demand as a guest speaker at the various learned societies, luncheon clubs and other gatherings which proliferated in and around Washington, New York and Boston.

Typical of Hudson's vigorous style are his talks to a Peace Congress Banquet in Boston in April 1907 and, a year later, to the William Lloyd Garrison Equal Rights Association of New York. Addressing the former on the subject of 'Arms for Peace', he was concerned to put over the message that in time of peace America had to keep its armed services in sound repair, even if only because it 'must, sooner or later, come into collision with Japan. While it is not a probable contingency in the near future, it is a possible contingency at any time. The United States is wholly unprepared for an attack in the Pacific . . . the Japanese, had they only the money, are amply prepared to . . . sweep us from the Pacific . . . and it would cost a lot of time and trouble to dislodge them.' As for the peace movement, its worst enemies were 'those hyper-sentimental friends who demand the abolition of armaments and who would brand the soldier a criminal . . . We need forts and guns along our shores, and battleships and cruisers to patrol the seas, just as we need bolts and bars upon our doors, and policemen in our streets.'

To the cause of equal rights for women Hudson was an early convert, won over by his devotion to Lilian and by the very real contribution she was making to the running of their affairs. Soon after they were married, he told his audience, 'there came one of those low-down financial depressions. I was living in a place in London, the rent of which was nearly $2000 a year, and our available cash amounted to but $19.' But their difficulties were overcome thanks to the moral and practical support of his wife, whom henceforward he treated as his equal in every way. Hudson was influenced by Lilian in another respect. Not the least of her attractions lay in the fact that she was a cultivated woman, having acquired from her father a genuine love of art, music and literature, especially poetry. Among Hudson's papers is a folder of verse written by Lilian in the manner of Robert Browning, and from about 1900 he also tried his hand at occasional prose pieces or poems to commemorate events such as the San Francisco earthquake or the death of a valued friend.

After about 1907 the couple started the custom of holding 'creative parties', the spirit of which is conveyed in an invitation addressed to a selection of friends among whom were obviously a number of military and naval officers. The guests were exhorted to:

. . . meet and mix with laughter, mingle with the song and hold parley with the erasing years. At our house, 698 St. Marks Avenue, Brooklyn, there is to be a round-up of genius, science, art and literature, with representatives and ambassadors of Mars and Neptune. Bacchus and Ceres are to do the catering. The horn of plenty will be found upon the old oaken board at half-past six, after which the Nine Muses will recite, divine Cecilia sing, Bacchante dance and Pegasus be ridden bareback. Hours six to twelve.

On these occasions Hudson enjoyed entertaining the company with sensational anecdotes about his adventures with explosives at a time when safety regulations were virtually non-existent and disaster never far away. So popular did these prove that he was prompted to write them down, though not until 1916 did his collection of *Dynamite Stories* find a publisher.

By now Hudson was on the closest possible terms with the Durban family. He and Lilian exchanged long, affectionate letters with Dr and Mrs Durban and their other two daughters, and in 1908 it was decided that young Will should travel from England to Hopatcong to take over the management of the Hotel Durban. This turned out to be an unsatisfactory arrangement, Will having to be formally reprimanded for neglect of duty, smoking cigarettes (in Hudson's view a serious misdemeanour) and making improper advances to one of the secretaries. More productive was a correspondence between Lilian and Hudson on the one hand and Dr Durban on the other, during which Hudson formed the opinion that 'modern verse has degenerated largely into twaddle'. He was therefore moved to write a book which set out systematically to apply scientific method to the study of literature. 'Poetry and gunpowder,' he suggested, 'were both invented about the same time . . . one of the first uses of poetry was the writing of the Bible, and since then gunpowder has been used mainly to back up that poetry.' The typescript was first submitted to Harpers, whose chief reader rejected it on the grounds that the author had 'no ear for verse . . . little sense of style, [and was] deficient in taste and literary culture'. Enraged, Hudson responded with a 32-page diatribe defending his work before sending the script to Funk and Wagnalls, who published it in 1910 with the title *The Science of Poetry and the Philosophy of Language*.

Hudson was inordinately proud of this, his first major essay in print. Hundreds of copies of the book were mailed to friends and acquaintances on both sides of the Atlantic, but despite a vigorous promotional exercise sales were poor and reviews mixed, ranging from extravagant praise to downright abuse. The *Times Literary Supplement* was loftily equivocal:

> In the intervals of inventing death-dealing explosives, Mr Maxim has applied his vigorous intellect to the consideration of the nature of poetry; and one result is a definition which he prints under a portrait of himself in the frontispiece – 'Poetry is an expression of insensuous thought in sensuous terms by artistic trope'. Mr Maxim has certainly a taste for what is fine in literature, as may be seen by his collection of 'great poetic lines', and students who are not deterred by such fearsome new terms of art as potentry, tropetry, literatry . . . will find food for thought in his views and in his criticisms of other poetical critics.

Others were less polite. Writing in the *Book News Monthly* the young Ezra Pound was scornful, stinging Hudson to refer to 'that irresponsible tirade . . . purporting to be a review of my book.' To relieve his feelings the author took a whole page of the *New York Times Literary Review* in order to regale readers with the opposing and contradictory views of the critics. He also bound all the favourable comments into a pamphlet entitled 'The World's Verdict on Hudson Maxim's *The Science of Poetry* . . .: The Greatest Work of the Century on the Science of Literature'. 'The world's most eminent scientists, littérateurs, artists, poets and scholars,' he declared with the familiar Maxim hyperbole, 'have given the work their high approval and authoritative praise, hailing it as the most important treatise ever written upon the subject with which it deals.'

No doubt the less than favourable reception given to Hudson's book acted as balm to Hiram's soul, bruised as it had been by his brother's new-found prosperity. The older man's fortunes had not improved after the Boer War and with a return to the conditions of peace which prevailed during the 1890s. Shortly before his death the great Lord Salisbury is said to have remarked grimly to Hiram that he had saved more men from dying of old age than any man alive, but this achievement was not reflected in the marketplace. In 1903 and 1904 Vickers' Maxim Branch again lost money as the demand for guns and ammunition fell away, plant was closed down and workmen were laid off. The parent company having taken over the Wolseley Tool and Motor Car Company to provide an alternative source of revenue, Sigmund Loewe was tragically killed in a road accident while driving one of the new automobiles. It was therefore his successor as general manager, Lieutenant Trevor Dawson RN, who complained before the

Hiram shows his grandson Maxim Joubert how to fire the Maxim gun, c1910. (Science Museum/Science and Society Picture Library)

Hiram Maxim's Captive Flying Machine in action soon after opening on the South Shore, Blackpool. (Philip Jarrett)

In the South African War a Boer commando poses with battle-scarred Vickers 37mm 'pom-pom' and rifle calibre Maxim (left). (Imperial War Museum: Q101770)

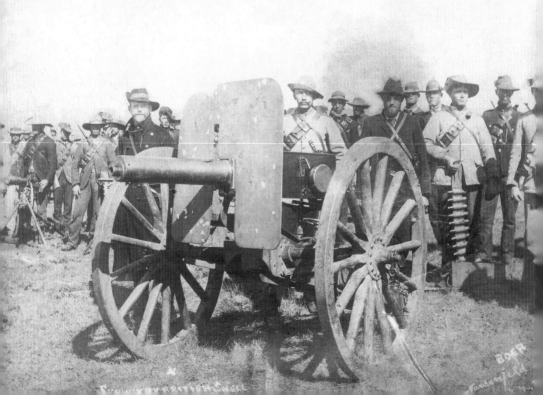

The Civilian Advisory Board, Washington, October 1915. Hudson stands front left between Edison and the young Franklin Delano Roosevelt, then Assistant Secretary of the Navy. (Hudson Maxim Papers)

German machine gun section with the DWM adaptation of the Maxim gun which accounted for hundreds of thousands of Allied soldiers during the First World War. (Imperial War Museum: Q23709)

Hiram at the drawing board: his favourite occupation. (Philip Jarrett)

Left: The grey granite memorial to Hiram and Sarah Maxim and 'the Boy' in West Norwood Cemetery. (Author) *Right*: Hudson Maxim's plaque at the former entrance to the Morris Canal, Lake Hopatcong, New Jersey. (Author)

Royal Commission set up to investigate the conduct of the war in South Africa that since to meet the emergency Vickers had been obliged to divert much of their trade to Germany, business had been lost which could not easily be regained. In consequence, he declared, 'a large part of the Maxim guns are now being manufactured abroad instead of at our works . . . and we hope now that the war is over that the sacrifices we have made will be considered.'

But his plea fell on deaf ears. Vickers were now concentrating on the lucrative business of turning out warships and heavy ordnance for the British as well as foreign navies, compared with which the manufacture of machine guns was of minor importance. With orders from the War Office reduced to a trickle, production of Maxim guns in England dropped to less than 100 a year, or about one-fifth of capacity, so that at Erith nearly half the machines were standing idle and Crayford was closed for want of work. Hiram had little to do other than brood over his grievances and especially the depredations caused by his brother,[4] give the occasional talk to learned societies and potter in his laboratory at Thurlow Lodge. Although a regular attender at board meetings, he was increasingly honoured in deference to his reputation rather than for any other reason. Already he had sold most of his shares in the company and been obliged, in a curious echo of his agreement with the electrical consortium, not to carry out any more work on machine guns. Nevertheless, in December 1902 he was confirmed as 'Consulting Engineer and Special Director', charged with producing inventions 'as the Vickers Company may from time to time require but not otherwise,' in return for a basic director's salary of £1200 a year.

Hiram's health was now giving cause for concern, subject as he was to recurrent bouts of bronchitis brought on by the cold and foggy English weather. The treatment prescribed by leading British specialists proving unhelpful, he was recommended to a clinic at Le Mont Dore, near Vichy, where he 'submitted to a very long system of steaming and boiling and taking the waters.' This having no effect, it was suggested that he resort to the Inhalatorium run by a Mr Vos at Nice in the south of France. Accordingly Sir Hiram and Lady Maxim spent the winter months of 1902-03 partly in Nice and partly in Monte Carlo, where they stayed at Basil Zaharoff's favourite Metropole Hotel. The course of treatment at the Inhalatorium was thorough and

---

4. In 1904 he wrote to his old schoolteacher in Maine that Hudson 'had by his skilful writing and shameless falsehoods made at least half the American Public believe that he is the real Maxim and I am only the bogus imitation'.

severe, involving daily inhaling sessions lasting an hour at a time, but it was successful. By April the bronchitis had disappeared and Hiram returned to London, only to be struck down again in the following autumn. There was nothing for it but to go back to Mr Vos, but this time Hiram made sure that he found out exactly how the inhaling process worked so that he could adapt it for use in England.

For someone of Hiram's temperament these long sojourns in Nice would have spelled boredom had he not found another outlet for his limitless curiosity. As on an earlier visit to Monte Carlo, he was drawn irresistibly to the Casino, where observation of the play convinced him that gambling was a mug's game, since mathematically the odds were always loaded in favour of the Bank. In a letter to the Paris office of the *New York Herald*, Hiram quoted the proprietor of the Casino, Monsieur Blanc, who was reported as saying in an unguarded moment: 'Rouge gagne quelquefois, Noir souvent, mais Blanc toujours', and he attacked Mr Herbert Vivian's assertion that by means of an 'evening up' system the Bank could be confounded. This sparked a long and lively debate. In a rejoinder headed 'A Small-bore Maxim', Vivian took his stand on the law of probability, declaring that 'Sir Hiram Maxim . . . has shown himself to be an unpractical and illogical person . . . If he directed all his affairs in a similar spirit, he would not be fit to remain at large. Were I a Yankee, I should exclaim "Hiram makes me Tyred!" '

Early in 1903 the dispute spilled over into the British press, *The Times* approving Hiram's letter as 'a singularly impressive warning against an unprofitable form of investment . . . which cannot fail to prove convincing to all except the incorrigible'. Support also came from the Irish MP and newspaper owner T P O'Connor, while Hiram was opposed by Lord Rosslyn and others. Although conducted throughout with tongue in cheek, the controversy aroused enough interest for Hiram to write it up in his first book, *Monte Carlo Facts and Fallacies*, which was published in 1904. Illustrated by 'Character Sketches made in the Casino by Sir Hiram Maxim', this confirmed the author's response to would-be gamblers – 'Don't!' He had, however, no expectation that his advice would be taken: people, after all, went to Monte Carlo not to make money but to enjoy the thrill of playing against the odds, and they did not mind paying for the privilege. It was with this in mind that he ended by quoting a quatraine printed in the *Pall Mall Gazette*:

> Don't gamble!' Sage Sir Hiram cries –
> But, though such sound advice is needed,

When shouts of 'Faites vos jeux!' arise
The wisest Maxim goes unheeded![5]

Returning to England in the spring of 1904, Hiram sought to ward off further attacks of bronchitis by inventing his own pocket medical inhaler, which he always referred to as his 'Pipe of Peace'. This made use not of Mr Vos's cocaine but of an infusion of his mother's favourite plant lobelium with pine essence. 'By making a mouthpiece of such a shape that the vapours were introduced directly into the throat,' he wrote, '. . . I found that my simple device was much more effective than the very elaborate machinery of Mr Vos.' So successful was it, indeed, that some years later he was to arrange for the inhaler to be manufactured by the firm of John Morgan Richards and Sons of London, since when it has helped to relieve sufferers all over the world.

More immediately, Hiram was encouraged to revive his interest in the aeroplane and the possibilities of manned flight. In 1902 he had been invited by E Seton Valentine and F L Tomlinson to contribute a preface to their *Travels in Space, a History of Aerial Navigation*, in which he repeated his view that despite the spectacular achievements of Count Zeppelin in Germany, lighter-than-air machines were reaching the limit of their capability. The prospects for heavier-than-air machines, on the other hand, were improving as designers learned from experience and progress was made with more efficient petrol engines developed for the automobile industry, not least by Vickers' subsidiary the Wolseley Motor Car Company. In the light of recent researches, and particularly his own, Hiram had every confidence that: 'Propulsion and lifting are solved: the rest is a matter of time.'

One problem was that flying experiments were apt to be expensive. In France and Germany inventors had a reasonable expectation of financial support from official sources, and after the Spanish-American war the US War Department had provided Professor Langley with $50,000 to pursue his experiments with manned aircraft. Unfortunately the British government showed no sign of following this example, nor were Vickers prepared to help. However, during 1903 Hiram took a lively interest in trials with man-lifting boxkites being conducted at the Crystal Palace by the American Samuel Franklin Cody. In June the Aeronautical Society held an International Kite Competition on the Sussex Downs, near Findon. Both Major Baden-Powell and Cody were competitors, and Maxim was on the jury.

5. In 1908 Hiram accepted a challenge from Lord Rosslyn, who was determined to prove his skill at roulette. The contest took place in Hiram's rooms in Kensington, Rosslyn's system failing when after 3080 coups he had lost all his money.

A former Indian fighter, cowboy and gold prospector, Cody had, like his friend Buffalo Bill, toured England with a Wild West riding and shooting act, and it may have been he who suggested to Hiram the idea of subsidising further experiments in aviation by converting his 'whirling arm' apparatus into a fairground attraction. This Hiram mulled over during his stay in the south of France, and on his return he discussed the prospects with people experienced in the entertainment business, who confirmed that the machines could be profitable if erected at such popular resorts as Earl's Court, Blackpool and the Crystal Palace. Accordingly a small company was formed, and Hiram quickly produced the working drawings.

The original 'Captive Flying Machine', as it came to be called, was built by Hiram at West Norwood with the assistance of the talented young aero-engineer Albert Thurston. It consisted of a vertical tower from which angled arms dangled outwards and downwards with cigar-shaped cars attached to their ends by wire rope. The shaft was turned by a petrol engine which caused the cars, equipped with wings and a propeller at the rear, to fly outwards and upwards by centrifugal force, giving passengers the sensation of flying. In July 1903 the first patent was filed for 'Improvements relating to rotating cars or roundabouts for public recreation or experimental purposes . . .', and soon after the prototype was ready. Thurston was much impressed by Hiram, whom he described as 'always teeming with ideas and inventions', and he particularly admired the spirit which drove his chief to restless activity in the cause of aviation. Many years later he recalled how as they were toiling away at Thurlow Lodge one pleasant sunny afternoon:

> He was on one side of a wing, I was on the other and there were one or two workmen present; suddenly the wind brought the sound of Chopin's Funeral March from Norwood High Street. Maxim nodded in the direction of the cemetery and said: 'Ah, boys! One of these days when I'm over there, you will think of me working like hell-fire and damnation (a pet phrase of his) for an impossible ideal which will revolutionise the world.'[6]

Alas, however, the project was to founder in a sea of administrative and legal troubles, and to end up losing money. Any notion that the passengers in the cars might operate the wings as they flew round the circle was firmly (and no doubt wisely) rejected by the inspectors of the London County Council on grounds of safety. Without this

---

6. A P Thurston, *Reminiscences of Early Aviation*. Presidential address to the Newcomen Society, Science Museum, London, October 1949.

feature the machine became, in Hiram's words, 'simply a glorified merry-go-round'. Still, in an item headed 'Sir Hiram Maxim's Captive Flying Machine', *The Times* reported on 18 March 1904 that the device, featuring radial arms with slung 'boats' each carrying six or eight passengers, was to be installed fronting the lake at Earl's Court, where there was enough space to show off the aeroplane effect to best advantage. Although an even larger machine was in prospect for the Crystal Palace, the main object was not to amuse but to defray the costs of serious experiments in aeronautics, for the inventor believed that it was now practicable 'to make a flying machine that cannot fail to be of enormous value to the country as a military engine'.

Initially the Earl's Court machine fulfilled all expectations, taking £325 on its first day in operation. By the end of the season it had, despite the occasional breakdown, earned £8000, and Hiram and his fellow directors were encouraged to press ahead with additional machines at Blackpool, Southport and the Crystal Palace, which were built by the Maxim Electrical and Engineering Company of London. There were, however, many pitfalls along the way. Delays in construction more than doubled the cost of erection at Southport and Blackpool, and there were constant difficulties with the larger and more ambitious machine at the Crystal Palace. Soon Hiram found himself in deep water financially and involved in legal entanglements with shareholders and commercial sharks. Frustrated and exasperated in equal measure, he decided to settle out of court and wash his hands of the whole affair, with the result that by his own admission he lost over £10,500 on the enterprise, 'and nobody made any money out of it except the lawyers and one of the promoters'.

Just the same the captive flying machines went on entertaining the crowds. In 1911 it was reported that at Southport and Earls's Court they were a source of delight to visitors, who were provided with 'a thrilling and exhilarating ride through the air' at up to 40 mph. As to the machine at the Crystal Palace, this continued in action until 1936, when it was badly damaged in the great fire; it was finally dismantled in 1949, when the '*600*' magazine referred to it as 'The Wonder of the Age'. The fourth machine, extensively renovated, is still in use at the Blackpool pleasure beach, and is the only working example of the inventor's art to be seen anywhere outside of a museum.

While the failure of this money raising venture was a disappointment, it did not diminish Hiram's interest in aeronautics, and he continued to be consulted by pioneers such as S F Cody, who during the next few years was to become one of the most successful of British aviators. Through H G Wells he met Lieutenant John W Dunne (later to achieve fame as the author of *An Experiment with Time*), who, having been invalided out of the South African War, was trying his hand at designing an inherently stable tailless biplane. Dunne's researches included a visit to the Crystal Palace, where he was whirled around at high speed on an apparatus used by Hiram during the off-season for testing aerodynamic devices, the passenger cars being removed and replaced by a single rotating arm. Hiram offered him a job as experimental assistant, but instead Dunne joined the Balloon Factory at Farnborough, where he designed a biplane glider 'which when it was suspended from a kind of revolving gallows at the Crystal Palace attained a speed . . . of seventy miles an hour and rose to a height of seventy feet'.[7]

Fortunately for Vickers, new commercial opportunities were opened up by the Russo-Japanese War, which broke out in February 1904 and lasted until August 1905. For the first time in a major clash between industrial nations, artillery and small arms using the latest smokeless propellants and high explosives were tried in action. Since neither Russia nor Japan had the technical resources to produce these materials for themselves, they were largely supplied through the agency of British, French and German armament companies, a process which was fraught with political sensitivities. From the outset British public opinion strongly favoured the Japanese, anti-Russian feeling being exacerbated by the Dogger Bank incident, when the Baltic Fleet on its way to the far east fired on British trawlers, mistaking them for Japanese torpedo boats. However, under the terms of the Anglo-Japanese treaty of 1902 Britain was obliged to remain neutral, and so any assistance given to either side had to be unofficial and indirect.

In the event substantial help was given to the Japanese, and particularly their navy, through more or less clandestine channels by Vickers and Armstrong-Whitworth, who collaborated in order to compete with their foreign rivals and take full advantage of the situation. The British companies also contrived to deliver submarines, torpedoes, shells and other war material to Russia by using a French-based intermediary, the Société des Munitions Françaises, set up for the purpose by Basil Zaharoff. As to machine guns, the Russian war department made use of French bank loans to import hundreds of Maxims via Warsaw from Deutsche Waffen of Berlin. These weapons,

---

7. Walter Raleigh, *The War in the Air*, Vol I (Oxford 1922), pp114-5. Quoted by Philip Jarrett in his article 'Maxim's Device', *WW1 Aero* (February 1994). Dunne went on to develop amid much secrecy a powered biplane with swept-back wings which had some success.

on which Vickers continued to earn royalties, were deployed to good effect during the fighting in Manchuria, as were the Hotchkiss guns produced in Japan with technical and financial assistance from the French. Only after the war were the Russians able to begin the manufacture of Maxim guns for themselves at the Tula Arsenal in St Petersburg.

As a result Vickers' Maxim Branch, having lost money in 1903 and 1904, showed a temporary profit of £107,000 in 1905 and £233,000 in 1906, largely from overseas sales and royalty payments deriving from Germany and Russia. Since Hiram had ceased to have any direct financial interest in his gun, he did not benefit from this sudden increase in demand to anything like the same extent as his friend Zaharoff (who in 1906 was paid a staggering £29,000 by way of commission and 'agency expenses', a euphemism for bribes). By now he was for most practical purposes sidelined within the company, and, having burned his fingers with the captive flying machine, he opted for a quiet life, drawing his director's salary, seeking to market his medical inhaler and patenting an improved apparatus for vacuum cleaning and the extraction of dust. He also began, with Sarah's help, to draft his major work on the science of aviation.

In 1906 Hiram was saddened to hear of the death in the United States of his younger daughter Adelaide. Some six years earlier she had married Eldon Joubert, a musician who earned a living as an impresario and piano tuner in New York City, but the relationship was not a happy one and Adelaide appears to have gone into a decline. The couple had one son, Maxim Joubert, who it was arranged should come over to live with Hiram and Sarah in England. In 1907 the boy made the Atlantic crossing, afterwards settling down and evidently getting on well with the famous, if somewhat intimidating, grandparents he had never seen. At long last Hiram had a surrogate son whose presence went some way to compensate for his alienation from Hiram Percy, and on whom he was able to lavish such affection as he had to give. Always referred to by Hiram and Sarah as 'the Boy', Maxim Joubert went to Dulwich College, becoming a noted athlete and in due course serving with distinction in the American army during the Second World War.

SEVEN

Marching as to War

By increasing the horrors of war, maybe,
You have carried us nearer to peace.
So feel no remorse, that so fearful a force
As your gun has caused thousands' decease.

I don't really see why the world should complain,
If they hadn't died thus they'd have died all the same.
Therefore on your birthday be merry and bright,
But please drop the problem of aerial flight.

This doggerel verse, addressed to Sir Hiram Maxim on the occasion of his seventy-first birthday (and also of his retirement from Vickers), appeared on 5 February 1911 in the 'Birthdays of the Week' column of the popular magazine *John Bull*. It is typical of the *sang froid* with which press and public at the time generally regarded the latest advances in military science. Although during the previous twenty years armies and navies had developed greater destructive power than during the whole of the preceding two centuries, at the time the awesome nature of the changes taking place was not immediately apparent: if it had been the prospect of hostilities would not have been greeted in the summer of 1914 with quite such widespread enthusiasm.

It must be counted fortunate for Hiram and Hudson Maxim that their careers happened to coincide with the beginnings of what is now known as the military-industrial complex, that formidable alliance of interests founded on applied science, private enterprise and the armed services. Although anxieties about the international arms race found expression at the Hague Peace Conferences of 1899 and 1907 (when the word 'pacifist' was coined), there was little serious opposition to the growing dominance of the armaments industry. Prophets of doom invoked the spectre of war, and pressure groups published pamphlets condemning the armament rings which were allegedly engaged in fomenting it. Nevertheless, the theory of deterrence was widely accepted and the Maxim brothers, in common with arms manufacturers like Alfred Nobel and Andrew Carnegie, believed that they were the true pacifists and that only by means of a balance of armed might could the nations protect themselves against aggression, and so make wars impossible.

It is difficult to determine with any certainty how far Hiram's later researches influenced the initiatives taken by the Vickers company. Apart from automatic guns, his contribution to the on-going development of ordnance, projectiles and explosives was considerable. In other respects it was marginal, as in the case of his work on automobiles and lightweight petrol engines, which may, just the same, have played some part in Vickers' decision to acquire the Wolseley Motor Car Company. Certainly Hiram was concerned to explore the possible military uses of motor vehicles;[1] the first designs were for high-speed cars for staff officers and scouting duties, and armoured cars were contemplated from the beginning. Since the products of the Wolseley factory were built to a high standard, they were soon in demand as luxury cars for civilians and by 1913 were selling in large numbers.

In the matter of aviation Hiram's contribution was a minor one, though not without significance. The episode of the captive flying machines did little to enhance his reputation either with Vickers or officialdom, but he continued to be recognised as an expert and consulted by the press on anything to do with aeronautics. Since he remained convinced that the aeroplane could not be expected to compete as a carrier with railways and steamships, Hiram saw its only practical potential as a weapon of war, and in this regard he found himself at variance with the military and naval authorities, who during the early years of the century were inclined to put their faith in dirigible airships. This preference was hardly surprising given the rate of accidents and failures with heavier-than-air machines. Not until 1907 did any European aviator succeed in raising himself aloft for more than a few seconds, and in that year R B Haldane, Secretary of State for War, echoed the general view when he declared that 'aeroplanes will never fly'.

Nor at first did the Wright brothers in America fare much better. Concerned to safeguard the secret of their mastery of controlled flight in the expectation of large military contracts, they met only with disappointment. In 1904 Colonel John Capper, soon to become superintendent of the army Balloon Factory at Farnborough,[2] went to North Carolina to visit the Wrights, recommending on his return that the War Office consider acquiring their aeroplane as a potential 'accessory of warfare'. The Treasury did not accept his recommendation,

---

1. On which subject he had dealings with the noted automobile engineer F R Sims, who was instrumental in introducing Gottlieb Daimler's engines into Britain.
2. Forerunner of the present Royal Aircraft Establishment.

nor did the War Office respond when two years later the Wrights offered their *Flyer* to the British government for a down payment of $200,000. Instead the Balloon Factory set Cody to work on a powered non-rigid airship, the *Nulli Secundus*. This, the first of a series of British military airships, made her maiden flight in October 1907, when 'she danced, like an elegant little whale, round the dome of St Paul's Cathedral, and over the grounds of Buckingham Palace'.[3]

In recognition of this success the War Office made Cody a grant of £50 towards his work on a military biplane, but after he had achieved the first official British recorded flight (of 496 yards) in October 1908 his initiative was not followed up. It was, therefore, mainly the revelatory demonstrations of flying by Orville Wright in the United States and Wilbur Wright in France which led to a major breakthrough. The efforts of the former were rewarded when following army trials, and despite a serious accident which injured Orville and killed his passenger, the United States government at last agreed to buy the manufacturing rights to the Wright aeroplane. As for Wilbur, he treated admiring crowds to a series of ever more assured performances, breaking all records and in December 1908 winning the Michelin Cup with a flight lasting two hours and twenty minutes. Having earlier been sceptical of the claims made by and for the Wrights, Hiram was converted when in August he journeyed to Le Mans to see for himself the intrepid aviator in action. *The Times* was quick to note the connection, commenting apropos the pace of progress that: 'It is less than fifteen years since the first important experiments with a machine heavier than air were made in Kent by Sir Hiram Maxim.'

But the inventor was not on his best behaviour. Wilbur wrote to his brother: 'Maxim was here several days this week. I doubt the goodness of his purpose and dislike his personality. He is an awful blow [*sic*] and abuses his brother and his son scandalously'.[4] Hiram was indeed showing signs of his advancing years and the effects of deafness and indifferent health. He was, however, greatly impressed by his fellow countryman's prowess in the air, writing shortly afterwards that he 'had as complete control of his machine as a skilful boatman would have on a placid stream'. Back in England he added a postscript to his book *Artificial and Natural Flight* affirming that 'the remarkable success of the Wright Brothers has placed the true flying machine in a new category. It can no longer be ranked with the philosopher's stone . . . success is assured.' He was also moved to go back to the drawing

3. J D Scott, *Vickers, a History*, p70.
4. Harald Penrose, *British Aviation*.

board, redesigning and later in the year patenting an updated, more compact version of his earlier biplane, 'having pneumatic sprung undercarriage, with front and rear elevators'.

At a meeting of the Aeronautical Society a few months later, Hiram was not on his usual voluble form, and when he did speak he admitted that 'on account of my deafness it was quite impossible for me to understand what was said here tonight, though I caught a few words.' Confirming that he had developed a lightweight petrol engine and was well ahead with the design of a new machine, he repeated that the future of aerial navigation lay not with dirigible balloons but with aeroplanes, which would sooner rather than later be essential for the defence of the nation and the Empire. This point he had already made in his book, published the previous autumn and now going into a second edition. The authorities, he wrote, would be well advised to recognise the dangers inherent in the new era of aviation, and particularly the threat posed by the Kaiser's Germany:

> The value of a successful flying machine, when considered from a purely military standpoint, cannot be over-estimated. The flying machine has come, and come to stay, and whether we like it or not, it is a problem that must be taken into serious consideration. If we are laggards we shall unquestionably be left behind, with a strong probability that before many years have passed . . . we shall have to change the colouring of our school maps.

It was a salutary warning, which fell on receptive ears. Hitherto, secure under the watchful protection of the navy, the British had, unlike their French and German neighbours, attached little priority to military aircraft, regarding aviation more in the nature of a sport or as an opportunity for displays of derring-do. Now the danger from Germany had become real, and in some quarters it was suggested that the Kaiser was deliberately creating a large aerial fleet in order to counter the overwhelming might of the Royal Navy. It was yet another reason for a national concern already reflected in the Anglo-German naval race and in heated debates in Parliament on the naval estimates, and helped to fuel the invasion scares which were obsessing press and public alike.

Early in January 1909 Winston Churchill, who took a personal as well as a professional interest in the possibilities of flying, invited Hiram to a meeting to sound out his views on aerial warfare. During their conversation the inventor stressed the versatility of aeroplanes as compared with airships. He put it to Churchill that the success of the

Wright *Flyer* was due mainly to the brothers' skill in flying, and that very soon more reliable machines, such as that on which he was currently working, would be available. With better machines and better pilots would come the capability for offensive operations, among which Hiram identified the bombing of naval bases and attacks on warships by means of high explosives lowered on piano wire. Intrigued, Churchill called for 'a very searching and authoritative investigation', and he urged Haldane to consult Hiram before the next meeting of the Committee of Imperial Defence.[5]

When the committee met a few weeks later, Hiram was summoned to give evidence before an Aerial Navigation Sub-Committee appointed to advise on the best way forward. In his usual flamboyant style he declared that since seeing Wilbur Wright fly at Le Mans he had revised his opinion on the time needed to develop the potential of flying machines. He now had no doubt that within a few years fleets of aeroplanes would be able to deliver bombs to targets many miles distant, and even act as aerial ferries to convey an invading army across the Channel. In the course of questioning, however, it became evident that what Hiram was really after was financial backing for his own aeroplane, and the members of the Committee were wary of what they saw as special pleading. Although a number of small firms were competing to come up with a practical aeroplane, few had succeeded in getting a machine into the air and keeping it there. On the other hand, although the French were doing better, both they and, significantly, the Germans were building fleets of dirigible airships which from the standpoint of the Admiralty appeared to be of greater utility than the aeroplane.

Accordingly the Committee's report came down in favour of airships, declaring that there 'appears to be no necessity for the Government to continue experiments in aeroplanes, provided that advantage is taken of private enterprise in this form of aviation.' Asked in the House of Commons whether the government intended to vote 'a substantial sum' for aeroplane construction, Haldane replied that it was proposed rather to rely on the initiative of private inventors. The War Office decided to halt further expenditure on flying machine experiments other than at Farnborough, and the Committee of Imperial Defence recommended to the Cabinet that £35,000 be included in the naval estimates to develop an airship of the rigid, or Zeppelin,

---

5. Alfred Gollin, *No Longer an Island*, pp422-6. Although at this time Churchill appeared to be in favour of airships, he was later to assert that he 'rated the Zeppelin much lower as a weapon of war than almost anyone else . . . and confined the naval construction of airships to purely experimental limits.' (*The World Crisis*, Vol I, 1911-14, p313).

type. Already Vickers had submitted plans for the construction of such an airship, and these were now approved. At once the company put in hand the building of No 1 Rigid Naval Airship at Barrow, a project that went ahead at the expense of any further effort devoted to heavier-than-air craft. To Hiram personally this was a cause of no little annoyance: once again he was obliged to subsidise the work on his aeroplane from his own pocket, although the company did agree to his using the facilities at Crayford together with a test site at Joyce Green, near Dartford.

No sooner had the authorities opted for the airship than the advances made in heavier-than-air technology began to show more positive results. During 1909 the French and German war ministries completed their purchase of the rights to manufacture aeroplanes on the pattern of the Wright *Flyer*, and on Laffan's Plain at Farnborough Cody's biplane, the so-called *Flying Cathedral*, achieved flights of ever longer duration.[6] In July Louis Blériot crossed the Channel, the first of a succession of aviators who by so doing were to bear out the fears expressed by Hiram and bring home to the British people that their supposedly impregnable island was suddenly become more vulnerable. In August a 'Grande Semaine d'Aviation' was held near Reims at which generous prizes were offered for races and exhibitions of aerobatics. The opening was attended by Hiram as well as notables and journalists from all parts of the world, and during the week it was announced that work was well advanced on his latest, all-metal machine which was being built under the supervision of A P Thurston[7] and equipped with a four-cylinder Wolseley petrol engine.

The Reims meeting served to underline the superior skills of the French pilots, who carried off most of the prizes, and in October the Paris Aero Show further indicated how far the continentals were extending their lead. Hiram took the opportunity to describe to reporters how before long the French would be able to send machines over London carrying a 400lb load in addition to fuel and pilot, and shortly afterwards Colonel F G Stone addressed the Aeronautical Society on the subject of aerial bombardment. The implications of war in the air were explored, Cody making reference to the development of a bombing plane fitted with Maxim guns. It was noted that Britain was falling behind France and Germany in expenditure on aircraft of all kinds,

---

6. Cody continued to go it alone without official backing until 1913, when a new machine he was testing broke up in the air, killing him and his passenger.
7. In later life Thurston (1881-1964) became an authority on aerodynamics, specialising in the use of aluminium and other light alloys in aircraft construction, and a member of the Council of the Royal Aeronautical Society.

and that British aeroplane manufacturers were inhibited by fear of infringing the Wright brothers' patents. This may or may not have played a part in delaying the completion of the new machines on which Cody and Hiram were working, and which they hoped to display at the Aero and Motorboat Exhibition mounted at Olympia in March 1910 with the object of encouraging investment in the nascent British aircraft industry.

In the event, while many well-known names such as A V Roe, Handley Page and Short Brothers were represented, Hiram's three-seater pusher biplane was not ready in time. Two weeks earlier the machine had been displayed at the Crayford works, where it had been built with the help of Wolseley company technicians. Already it had attracted a good deal of publicity, as now did Hiram and his friend Cody as they passed among the exhibits at Olympia: 'All heads turned to follow the top-hatted and frock-coated figure of bearded Sir Hiram and the even more flamboyant Cody, with stiletto moustache and pointed beard, and wearing a splendid double cape and tweed hat.'[8] Hiram was, however, a worried man. As he wrote in an article in *Flight* magazine, the trials being conducted with his aeroplane on a large, tarred sand, circular track built at Joyce Green were seen as 'a very simple manner of teaching men to fly, because they can do it without danger to themselves or to the machine'. But already, either because this method of instruction was less than effective or because the design of the machine was fundamentally flawed, it was apparent that much more work had to be done.

Nevertheless, and with active support from Thurston, Hiram did his best to persuade his colleagues at Vickers that all that was needed was more time to make the necessary adjustments. As always it was the determination to perfect his machine that drove him on, though he sought to justify his efforts and the expense involved by more pragmatic, and more familiar, arguments, warning interviewers that: 'If a warlike continental nation should be the first to achieve success, it would probably make its power felt and rearrange things to suit its own ideas . . .' At the same time, and sounding a more optimistic note, he suggested that 'when all the great nations find out how to fly successfully, then there will be no more war between them, and the great armaments which now exist will happily become a thing of the past.'

To an extent Hiram's cause was helped by the growing infatuation of the public with anything to do with aeroplanes and aviators. During

8. Penrose, op cit, p216.

army manoeuvres on Salisbury Plain a Bristol biplane was used to good effect for spotting the movement of troops, impressing Churchill and prompting Colonel Capper to declare: 'The necessity has now arisen for every warlike nation to have a sufficient aerial fleet, armed and equipped for offensive warfare . . . to put out of action the enemy's aerial forces before it can carry out its proper role of reconnoitring over the enemy's country and attacking vital points of communication.' But the War Office, like the Admiralty, had yet to be persuaded of the practical value of aeroplanes, which, even when able to take off, could not fly fast or high enough to be really effective or indeed secure from ground fire. The tendency, shared by the Vickers board, was to wait for clear proof of consistent and improved performance before deciding to commit funds to any particular type of machine.

In March 1910 Hiram was one of a number of speakers at the inaugural meeting of the Aerial League of the British Empire, the purpose of which was to agitate for more public expenditure on airships and aeroplanes to match the rapid strides being taken by Germany, France and Japan. The fact that so many pioneers in the field were French, and so few British, was deplored, and the Duke of Argyll moved a resolution calling for a 'programme of active propaganda and the founding of a national institute of aeronautics'. Hiram informed the gathering that he had almost completed his machine, which, with an engine weighing only 2½lb per horse power, 'contained no single idea or detail borrowed from the French'. However, a few months later the Vickers board, finally losing patience with Hiram and his machine, decided to look elsewhere and, acting on the advice of the Admiralty, sought to obtain a manufacturing licence on another aeroplane of proven viability. Their choice fell on a two-seater monoplane designed by the French engineer and aviator Robert Esnault-Pelterie, and in Paris Basil Zaharoff successfully negotiated the purchase of the exclusive rights to produce his machine for sale in Britain and the colonies.

The REP machine offered the obvious advantage that it had not only flown but broken a number of speed and distance records. It was also fitted with a hydro-pneumatic spring buffer which effectively reduced the shock of landing, still a frequent cause of accidents and damage. For Vickers it was an altogether reasonable prospect, and Hiram's reaction was altogether unreasonable. The ageing inventor saw the decision to abandon his machine as the last straw, the culmination of a series of disagreements with the Vickers board. He had long been regarded by his colleagues with something like embarrassment, and it was only a matter of time before they went their separate ways. In October he announced to the press that although his aeroplane was nearly ready he was ceasing work on it because he needed to take a rest and because in his seventieth year he was too old to undertake the task of piloting it himself. While production of the first Vickers monoplane based on the REP model was begun at the Erith works, he cast around for financial backers for a fresh project, namely the construction of a large attack aeroplane designed to drop bombs.

In March 1911 Hiram confirmed his resignation as a director of Vickers, although his services as consulting engineer were retained.[9] This step, reported *The Times*, he had taken in order 'to join with Mr Claude Grahame-White and M. Blériot for the purpose of developing a new aeroplane to be used in the time of war, which . . . would reconnoitre the enemy's position' and be equipped to carry high explosive bombs. Grahame-White had acquired land at Hendon, north of London, where he was in process of opening the capital's first aerodrome. Hiram had already worked on the bombs and was about to patent a release mechanism together with a device to make them explode on or above the ground ('Improvements in and relating to bombs for use in connection with aeroplanes or flying machines'), and, at the inaugural Hendon Air Show in May, Grahame-White treated invited members of parliament and other VIPs to a convincing demonstration of aerial bombardment, dropping 100lb bags of flour on to the outline of a battleship painted on the grass.

But although Hiram's views on the threat from the air gave rise to a vigorous press debate,[10] the projected enterprise came to nothing. In commercial circles doubts persisted about the future of aeroplanes and there were, besides, differences of opinion between Hiram and Grahame-White. Under the pressure of events Hiram had become more than usually cantankerous and intemperate. He lost no opportunity to attack his brother in the British press, and in February 1910 Dr Durban sent a cutting to Hudson with the comment: 'He is acting like some assassin . . . do not let this worry you too much . . . Hiram *always* over does it. He is an insane hyperbolist.' That same year Hudson decided to augment his father-in-law's meagre income by paying him a regular pension, and in the summer he arranged for Dr and Mrs Durban to cross the Atlantic and spend several weeks at Lake Hopatcong. They were asked to bring with them one of Hiram's anti-

9. His resignation was deplored by some who feared that dropping 'the inventive genius of the concern' would depress the company's share price.
10. Some months earlier a similar display had been put on by Glenn Curtiss in the United States to demonstrate the vulnerability of Dreadnoughts to aerial attack.

bronchitis inhalers: 'I imagine,' wrote Hudson of his brother, 'he is pretty hard up and is willing to do most anything to earn a few cents.'

Dr Durban's visit was an unqualified success, coinciding as it did with the publication of Hudson's book on the science of poetry. On his return to England he kept up the work of surveillance, asking Lilian in one of his regular letters to her to let Hudson know that he was 'looking constantly and carefully at papers of all kinds . . . Hiram is either mad or bad, or *both*.' And after Hiram's resignation from Vickers he wrote with no little satisfaction to report that the company he proposed to set up with Grahame-White had failed to materialise 'as sufficient share capital could not be obtained'.

Hiram's temper had not been improved by reports from America of the close links established by Hudson with the army and navy, and the reputation his brother was enjoying in the United States as a specialist in explosives. By 1906 the Powder Trust headed by Du Ponts controlled the manufacture of much of the high explosive and all the military smokeless powder manufactured by private firms, and in Congress its monopoly position was challenged as breaching the Sherman Anti-Trust Act. Consequently in 1907 the US government started an action against Du Ponts, while Congress took steps to double the capacity of the official powder-making facilities at Picatinny and Indian Head. As intended, this reduced the level of orders going to Du Ponts, which was obliged to investigate alternative uses for its output of nitrocellulose.[11] The measure also reduced the royalties earned by Hudson, who nevertheless continued to draw his retainer and an income from sales of smokeless propellant and high explosives throughout the American continent. As the *New York Times* reported in a feature article in January 1909:

> Every time a big gun is fired in the American Army or Navy some of Hudson Maxim's smokeless powder is used. He has made enough explosives to blow all the navies of the world out of the water . . . He knows guncotton, nitroglycerine, cordite, maximite, motorite and dynamite as the average man knows collar buttons and neckties.

Meanwhile Hudson pressed ahead with his researches into torpedoes, torpedo boats, submarines and naval projectiles. Few of his ideas were actually taken up by the authorities, but those that were could be very profitable, notwithstanding that the inventor had assigned all development rights to the Du Pont corporation. Thus,

---

11. Such as the manufacture of artificial leather for the upholstery of motor cars for its subsidiary, General Motors.

having collaborated with the Bureau of Ordnance on an improved delayed-action fuse for armour-piercing shells, Hudson supervised lengthy trials with it at Indian Head and the Washington Navy Yard. In the process he submitted no fewer than eleven patent applications before the right to manufacture the fuse was purchased from Du Ponts by the US government in 1909-10. This was no mean achievement. As a senior captain affirmed when making his report to the Navy Department, 'there is no nation today which has succeeded in perfecting a device of this character, and this fact only emphasises the value that may be attached to the Maxim fuse'. Accordingly, he believed, a payment of $75,000 was reasonable on condition 'that Mr Maxim agrees not to reveal the secrets of this device as perfected . . . thus sacrificing any financial benefit that he might obtain by disposing of it to a foreign government.'[12]

It is unlikely that Hiram knew of this further success, which is perhaps as well given that he had earlier worked with Lieutenant Trevor Dawson and the ordnance technicians at Vickers on similar patterns of delayed-action fuse, notably one 'especially applicable to picric acid shells', patented in 1902-03. Nor in all probability was he aware of Hudson's continuing researches on high explosives, which after about 1910 concentrated on experiments designed to improve the stability of smokeless propellant and to find a safer alternative to Maximite as the burster in heavy projectiles. By this time Hudson was well established in the United States as a reliable specialist where any kind of explosives, whether civil or military, were involved, and he was regularly charging from a hundred to a hundred and fifty dollars a day for his services as consulting engineer or an expert in legal cases.

By contrast Hiram was gradually becoming more of an observer and commentator than an active participant in affairs. Soon after the arrival of 'the Boy' there was a further complication in the Maxim household when, according to a news item of 1911, the inventor 'took a fancy' to Vera Le Fleming, a lively and attractive young girl he saw playing in the grounds of the Crystal Palace with her nurse. Subsequently he and Lady Maxim decided to take the child 'under their wing', and she proved to be something of a prodigy as a mimic and entertainer, appearing in 1911 on the bill at the London Hippodrome, where she impressed audiences, including no less a personage than Queen Alexandra, as a 'clever impersonator with a varied repertoire'. No reference is made elsewhere to this curious episode, nor is anything known about Vera Le Fleming's later career.

---

12. Captain F Fletcher USN to Assistant Secretary of the Navy, 21 April 1910. National Archives, Washington.

Early in 1911 the Vickers company suffered a setback when the Admiralty, unimpressed by the claim that the new REP machine was capable of 'ascending from any of HM ships', informed the Board that it was 'not proposed to acquire any Aeroplanes for the Naval Service at present'. Only a few machines were therefore made, mainly for sale abroad, and a year later the British authorities banned the use of monoplanes following a number of serious accidents. Not long after-wards Hiram's biplane was badly damaged during taxiing trials at Joyce Green, and this finally brought to an end his long association with the business of aviation. By this time the company had dropped his name from its title, which became simply Vickers Ltd, though its guns and shells continued to be stamped with the initials VSM and his influence lingered on.

In September 1911 the reservations felt by many about the viability of airships were borne out when after many delays the *Mayfly*, the huge dirigible built by Vickers for the Admiralty, broke up in a strong wind while being brought out of its shed at Barrow and had to be abandoned. Hiram had always maintained that such craft could never be of real value either in peace or war since they were too fragile and too large to be manageable except in a dead calm, and the disaster drove home the message that the future of the air service lay with heavier-than-air craft. It was therefore on these that the Vickers com-pany technicians henceforward concentrated their efforts. At the Aero Show at Olympia in 1913 they exhibited the prototype of a new ma-chine which was to be used extensively during the early war years. This was a two-seater pusher fighter biplane, the FB1, equipped with an 80hp Wolseley engine and a Maxim-Vickers gun mounted in the nose and operated by the observer. At first output was limited by the unreliability of the power unit, but when the Wolseley engine was replaced by a 100hp Gnôme the FB2 became one of the world's first true fighter aircraft. Developed at the Crayford factory and tested at Joyce Green, successive versions were later turned out in numbers for the Royal Flying Corps. They came to be known collectively as the 'Gunbus', and the tradition of aeroplane building thus established was to be carried on with distinction into the post-war era.

Even as he parted company with Vickers, Hiram found himself embroiled in more controversy, this time with the American author-ities on the subject of smokeless propellants. A perennially explosive issue, this had smouldered on in military and naval circles until in the autumn of 1910 it again burst into flame, engaging the attention of service chiefs on both sides of the Atlantic and eventually the president of the United States. As we have seen, early doubts concerning the use

of the new propellants centred on the fearsome power of nitroglycerine, and particularly its corrosive effect on the barrels of heavy
naval guns. This was the reason why after 1901 the British modified
their cordite by reducing the proportion of nitroglycerine, and why
the US authorities chose to follow the example of the French and
Russians and opt for the single-based nitrocellulose powder from Du
Pont rather than the double-based cordite and ballistite produced by
the companies of the Anglo-German dynamite trust.

Carrying as it did obvious commercial implications, the debate
came to centre on the relative merits or demerits of the British and
American powders. British supporters of cordite argued that
nitrocellulose-based propellant, being liable to decompose and become unstable, was dangerous and could give rise to accidental explosions, especially in a hot climate.[13] The Americans, by contrast,
believed that the greater risk was presented by nitroglycerine, which
they had long declined to admit in any shape or form. They insisted
that nitrocellulose, if properly made to the necessary high standard,
was perfectly safe, and that by using the Maxim-Schupphaus multi-
perforated powder they were avoiding the problems of erosion caused
even by modified cordite in the guns of the Royal Navy. Underlying
the whole issue was Hiram's sense of injustice at the manner in which
his brother had, as he saw it, misappropriated the results of their joint
researches in order to profit from the sale of his powder to the US
authorities. Never a man to lie down under a grievance, he had over
the years bided his time and awaited his opportunity.

This came in May 1910 when an article in the journal *Engineering*
headed 'Nitro-cellulose Powder for Naval Purposes' drew attention to
the dangers of the propellant used in the French and American fleets.
It was noted that although three years earlier the French battleship
Jéna had blown up in the harbour at Toulon, the French authorities
were still persisting in the use of nitrocellulose rather than the double-
based propellants preferred by the British and Germans. This despite
the fact that the British authorities had 'carried out some few years ago
a most prolonged and exhaustive series of comparative trials', the
result of which was 'utterly to condemn' the nitrocellulose powder on
account of its instability. It was true that the French and American
navies were aware of the problem and had taken steps to guard against
it, but the danger remained. As to the tendency of cordite to erode the

---

13. According to Mottelay, the nitrocellulose first produced in Europe 'was badly made, and
    consequently very unstable. It was liable . . . to give off nitrous fumes, and to explode
    spontaneously, and it was this peculiarity that gave it a "black eye" in Europe'. Op cit, p112.

gun barrels, this had been exaggerated and was in any case offset by the fact that nitrocellulose-based propellant needed a much larger charge to obtain the same ballistic result.

The authors of this interesting piece were W H Maw and Alfred B Raworth, editors and publishers of *Engineering*. Long-standing admirers of Sir Hiram, they had been moved to indignation by his account of how, taking advantage of loopholes in the patent laws, Hudson had adapted his smokeless powder inventions and sold them to the Americans. Their own researches appeared to justify all of Hiram's allegations. In the following September a 12in gun on board the US battleship *Georgia* blew up during firing practice, and in October a further article appeared in *Engineering* declaring that: 'Artillery experts, both in this country and in the United States, are strongly of opinion that the accident in question, and similar ones which have occurred previously in the United States, were all due to no fault in the design or manufacture of the gun', but to the use of multi-perforated nitrocellulose propellant.

The article went on to take up the cudgels on Hiram's behalf, stating flatly that he was 'the first to manufacture and patent a powder consisting of the two violent explosives nitro-glycerine and true gun-cotton . . .', which came to take the form of cordite. Accidents with guns had been caused only by the variant of his powder adopted on the other side of the Atlantic, and were 'due altogether to the shape of the powder employed – a shape which Sir Hiram found dangerous long before it was selected by the United States.' Some years before Sir Andrew Noble had tested some of this powder sent out from America, and had expressed the opinion 'that it was an interesting variety, but liable to produce extremely high pressures'. From information since received from American commanding officers, moreover, 'it would appear that both the Army and the Navy in the United States are committed, against their wish, to the use of perforated blocks of smokeless powders for all calibres and velocities. These officers are convinced that the multiple perforations are a source of danger to gun and gunners when using large charges for high velocities.'

Emboldened by this support, Hiram resolved to take his case to the highest authority. On 25 October he wrote to President William H Taft enclosing a copy of the *Engineering* article and urging that the American military carry out precautionary tests. The allegation that multi-perforated propellant was putting heavy guns at risk was a serious one, and the Secretary to the President, Charles D Norton, copied Hiram's letter to the Secretaries of the Army and of the Navy, asking them to 'look into this matter carefully'. Early in December both replied, defending the existing service propellant. Captain Austin H Knight, president of the Special Board on Naval Ordnance, maintained that contrary to Hiram's assertion this was *less* likely than cordite to explode in the gun barrel. He acknowledged that there had been a number of accidents to guns since the introduction of smokeless powder, but maintained that these had occurred mainly during experiments at the Proving Ground and in older guns designed to use black or prismatic brown powder. As to the suggestion that the American services had more accidents than those of other countries, 'it is doubtful whether this is the case, as the well known policy of other Governments to keep such accidents from the public makes it difficult to institute a comparison'.

The Secretary of the Navy, George von L Meyer, concurred, submitting a list of explosions in naval guns which, like that on the *Georgia*, were mostly attributed to weakness or wear in the barrels. 'The broad statement of Sir Hiram Maxim that the gun accidents of the United States Navy are due to multi-perforated powder grain, therefore . . . is shown to be untrue.' Nor, contrary to what seemed to be Hiram's assumption, had nitroglycerine ever been contained in any propellant used by the Navy. Thus 'it can be stated with truth that [Sir Hiram] is in no sense the inventor of the type of smokeless powder developed by the US Navy at the Naval Torpedo Station, Newport, R I and used in the naval service since 1899.' In conclusion, Meyer thought it 'unwarranted to carry out further experiments as desired by Sir Hiram Maxim [since] his statements in relation to our smokeless powder are unworthy of serious consideration, except as . . . to the evil effects of their wide publicity on those unacquainted with the subject.'

These comments were duly forwarded to Hiram, who reacted with predictable indignation. Reports of the dispute having reached the press, he replied to the Secretaries in Washington referring to 'personal and abusive attacks upon myself in the papers, some of them with flaming headlines and language of the latest Billingsgate type. You have unwittingly done me a great injustice'. And he went on, changing the line of attack:

> You speak of my powders as . . . containing a certain proportion of nitro-glycerine. You describe your own powder as being a pure nitro-cellulose powder . . . and therefore that I am in no sense the inventor of the powder which you are using. But . . . if you had examined my British patent dated November 8th, 1888, you would have found the exact powder that you have described to the President as the United States powder, made in the same way . . . that is, a pure nitro-cellulose military powder, completely without nitro-glycerine; in a word, the identical thing.

As to multi-perforated powder, Hiram conceded that the American variety had succeeded better than most, 'probably due to the excellent quality of your nitrated cellulose', but he still maintained that 'the leading artillery experts in Europe . . . are all agreed that [it] is treacherous in a high degree'.

This letter Hiram despatched on 17 January 1911 together with a copy of his smokeless powder patent of November 1888, and in the next issue of *Engineering* another article by Maw and Raworth appeared on 'Naval Ordnance and Smokeless Powder'. Clearly at Hiram's instigation, the authors reproduced the list of explosions in American naval guns compiled in Washington, and repeated the assertion that these had been caused by multi-perforated smokeless powder. They also sought to justify from the official record Hiram's claim to be the true originator of progressive smokeless powders and the inventor of both the aerial torpedo, which led to the development of 'the first cordite ever made' and the only reliable delayed-action fuse yet devised for shells and torpedoes. 'We are,' they declared, 'fully aware that there has been a great deal of discussion as to who is entitled to the credit of inventing smokeless powder . . .' and particularly the type of powder used by the Americans. However, 'a casual glance at the Patent Office records is quite sufficient to put the matter . . . in its true light. Sir Hiram Maxim seems to us unquestionably to be the inventor of the class of powder used in the United States at the present moment. This, we believe, is disputed, but since new powders are always patented, there should be no difficulty in procuring evidence as to the subject if it exists.'

In Washington Hiram's letter was passed to the chief of the Bureau of Ordnance, who on 2 February wrote to the Secretary of the Navy pointing out that the specifications in the 1888 patent 'do not describe the powder in use by the US Army and Navy'. He also noted that Hiram had changed the grounds of his argument 'from the form of the grain to [the origin of] inventions'. Hudson is nowhere mentioned in the correspondence, though his presence lurks behind it and the part he had played was well known. But the American authorities had no wish to become involved with the widely-publicised quarrel between the brothers. The Bureau therefore contented itself with repeating that 'the assumption of Sir Hiram Maxim as to the cause of gun accidents was in error', and it recommended 'that further correspondence with Sir Hiram in regard to this matter be discontinued, as it appears to be unwarranted and undesirable'.[14]

---

14. Correspondence in the Records of the Bureau of Ordnance, National Archives, Washington.

There the matter rested, though it was by no means the end of the story. Doubtless Hiram felt a certain grim satisfaction when in September 1911 the French battleship *Liberté* blew up at her moorings in Toulon due to a propellant explosion followed by the detonation of melinite-filled shells, much as the *Jéna* had done four years earlier. Hudson expressed the opinion that both incidents were attributable to the faulty manufacture of nitrocellulose by the French state arsenals, and he assured the American authorities that such explosions could not occur with the powder made to the standard laid down by Du Pont and the US navy. Nevertheless, the signs are that after 1911 adjustments were made to the composition of the American propellant with a view to making it safer to use and store. The US navy also followed the example of the French in acting swiftly to prevent any further disasters: manufacturing processes were tightened up, older stocks of propellant called in, and as a consequence there were no more magazine explosions involving French or American ships.

On the other hand the fate of the French battleships tended to confirm the results of trials carried out by the British authorities, and to reinforce their belief that cordite was indeed the safer propellant. Although the Vickers-built Japanese battleship *Mikasa*, carrying British-made cordite, had also blown up in 1905 during celebrations to mark the great naval victory over the Russians at Tsushima, the cause could not be determined with any certainty. The result was a degree of complacency in the Admiralty which during the First World War was to be shattered by a series of spontaneous magazine explosions in warships as well as by the blowing-up of a number of Royal Navy vessels under enemy fire, including, most crucially, three of Admiral Beatty's battlecruisers at Jutland.

For the first time in his life Hiram now found himself chafing for want of an outlet for his still-restless energy. Although physically in decline, his intellectual curiosity was unabated, and he was regularly consulted on issues having to do with the defence of the nation. Thus he declared himself in favour of the building of a Channel tunnel, which he thought would bring 'incalculable benefits' while representing no kind of threat, since it could easily be closed against an invader. Even as Lady Maxim urged him to accept retirement and settle down to writing his memoirs, he was looking out for new challenges. One such appeared in April 1912 when the world's latest and largest ship, the White Star liner *Titanic*, sank in mid-Atlantic following a collision with an iceberg. Hiram was as stunned as everyone else by the suddenness and scale of the tragedy, and at once he looked to find some means of preventing anything similar happening in the future. While press and public concerned themselves with the human aspects of the catastrophe, he addressed himself to finding a practical solution to the problem of collisions at sea. Specifically, he wondered whether 'ships could be provided with what might appropriately be called a sixth sense, that would detect large objects in their immediate vicinity without the aid of a searchlight'.

Having discarded purely sonar solutions such as the reflective echo of fog sirens, Hiram once again turned for inspiration to the natural world. He had long been intrigued by the mystery of the flight of bats, and had conducted experiments from which he concluded that the creatures send out from their wings 'a series of pulsations or waves after the manner of sound waves' but of a frequency which acted like an illumination, enabling them to judge distances and catch their prey in the dark. Now he set out the result of these experiments together with detailed drawings of an apparatus for producing 'atmospheric vibrations'. This device, when placed on board ship, would emit low-frequency sound waves and receive the return signals on a large diaphragm, recording them by means of a rod and pencil making 'zig-zag lines' on a piece of paper. In this way, Hiram reasoned, it would be possible for a vessel to detect in advance the presence of large objects in its vicinity such as other ships and, of course, icebergs.

The inventor lost no time in writing up his ideas in a booklet which, with the title *A new system for preventing collisions at sea*, was published by Cassell and accorded a respectful review in the engineering supplement of *The Times*. Hiram's apparatus, this pointed out, was 'as yet in the experimental stage', and indeed it appears to have gone no further. In 1904 the German Hülsmeyer had been granted a British patent for his 'telemobiloscope', a maritime anti-collision device based on radio signals which he successfully demonstrated in Cologne but which was never taken up. Not until many years later was research into the sending and receiving of radio waves to lead to the development of radar, which, first suggested by Marconi in 1922, did not become a reality until the 1930s. In 1912 Hiram's solution was seen as no more practicable than any other hitherto tried, and the only reaction to it appears to have come from members of the Society for the Protection of Bats, in answer to whose remonstrations the inventor was obliged to add a note denying that his experiments had involved any kind of cruelty.

Hiram's reputation continued, therefore, to depend mainly on his rifle calibre machine gun, which during the years before 1914 entered into general use throughout the world. As is their wont the French went their own way, preferring to stay with the Hotchkiss gun, while the American services contented themselves with acquiring a token number of the 1904 model Vickers-Maxim guns produced by the Colt company. The small British army was as well supplied with machine guns as any other, but for tactical reasons neither they nor the French deployed them with much enthusiasm. After about 1907 the British general staff was increasingly inclined to follow the example of their colleagues on the other side of the Channel in accepting the doctrine of the offensive, and, since the machine gun was seen primarily as a defensive weapon, both armies continued to neglect its use, as for the same reason they neglected the use of heavy artillery.

This being so, it comes as no surprise to find that much the greater number of Maxim-type machine guns produced before 1914 were manufactured under licence in Germany and Russia. Hilaire Belloc's frequently quoted tag from his verse saga *The Modern Traveller*:

> Whatever happens, we have got
> the Maxim Gun, and they have not,

refers strictly to the sphere of colonial warfare.[15] In the wider context this was far from being the case. After the Russo-Japanese War, in which nearly half the battlefield casualties were officially credited to machine gun fire, the Russians spent large sums on re-equipping their demoralised armed services, and at the Tula Arsenal production was started of the 'Pulemet Maxim', one of the longest-lasting of the Maxim gun variants. German military observers of the land fighting in Manchuria, unlike their British and French counterparts, also took note of the effectiveness of machine guns, recording in the official history of the campaign that they were 'extraordinarily successful. In defence of entrenchments especially they had the most telling effect on the assailants at the moment of assault'.

As a result the German general staff called on Deutsche Waffen to develop an improved version of the basic Maxim model, and in 1908 the 7.9mm MG08, complete with Zeiss prismatic telescopic sights and an adjustable 'sled' mounting, went into production at the Berlin factory and the Spandau Arsenal. It was an immediate success, selling widely abroad as well as being supplied to units of the regular army and reserve on the basis of one machine gun company of six guns to each three-battalion infantry regiment. Although pro rata this allocation was the same as that of the British, the German army was by 1914

---

15. The later lines are less well known: 'We hanged and shot a few and then
The rest became devoted men.'

many times larger than the British, as was therefore its complement of guns. As time went by, moreover, the Germans formed an élite corps of specially trained field and fortress machine gun detachments in order to compensate for a relative shortage of manpower vis à vis the French and Russians. When war broke out they were able to put some 5000 guns in the field, more than any of their opponents with the possible exception of the Russians, who were, however, less well-organised to make the best use of them.

Thus before the First World War Vickers' Maxim Branch was sustained mainly by royalties from the manufacture of machine guns for the German and Russian armed services. Not to be out-done, the company had after 1901 taken steps to develop an im-proved model which was made lighter by the use of steel and aluminium in place of the existing brass casing. Passing through various stages, this resulted in the production from 1908 of the Vickers gun, which following successful trials was adopted by the British army in 1912. Most of these went for sale abroad. In each of the years leading up to 1914 only an average of eleven Vickers guns produced at Erith (apart from the Maxims which continued to be made at Enfield) were ordered by the War Office, which meant that plant had to operate well below capacity. Since when war came the allocation of weapons was still two per battalion, the British Expe-ditionary Force went into action with fewer than 200 Maxim and Vickers guns altogether. Within months, and even before both sides dug themselves in along the western front, the War Office had placed orders with Vickers for 1792 guns, of which by dint of hast-ily enlarging the shop floor at Erith and Crayford 1022 had been delivered by the following June.

But this is to anticipate. In 1913, apart from the usual troubles in the Balkans, the prospect of war seemed to have receded. At Thurlow Lodge Hiram pottered in his laboratory, working on various projects including a recurrent favourite, the production of a marketable coffee extract. He also found time to compile and publish a *Scrap-book* on the subject of Christianity for the enlightenment of the Chinese statesman Li Hung Chang. The inventor recorded with pride how in London he had regular discussions with his old friend, who 'paid me the great compliment of saying that I understood Chinese philosophy and re-ligion much better than anyone he had ever met who was not a Chi-nese'. His view of the Christian religion was, however, hardly unbiased, and the compilation took the form of a diatribe against both its historical record and its missionary work in China. *The Times Liter-ary Supplement* was not impressed:

Sir Hiram is of opinion that the 'mischievous propaganda' of the missionaries has done enormous harm in China and 'resulted in the loss of millions of lives'; and his object is to show the Chinaman that we are 'not all fools' and do not all believe 'the absurd doctrines' the missionaries teach. From this point of view he enlarges, with copious quotations . . .

But it was the need to guard against the menace of the warplane that continued to be Hiram's main preoccupation. In November 1913, addressing a meeting of the Scottish Aeronautical Society at Glasgow, he again warned of the dangers ahead, declaring that the implications of aerial warfare were so manifest as to be beyond computation:

One Dreadnought costs over a million pounds, and it is safe to say that for that sum five hundred large and powerful aeroplanes could be made. Supposing we were at war with a continental nation . . . each of the enemy's machines would be able to visit London and return twice in the darkness of one night, each time bringing and dropping half a ton of nitro-glycerine . . . How would London look after a week of that kind of treatment? There is only one way to protect ourselves against attack by aeroplanes, and that is by other aeroplanes.[16]

Meanwhile, Hudson was well on the way to becoming a pillar of the American establishment, commuting by road and rail from New York and Lake Hopatcong to government offices in Washington and naval and military installations along the east coast. While taking on a variety of speaking engagements and more or less profitable business commitments, he continued to act as consultant to Du Ponts, who in September 1911 entered into another agreement confirming their interest in his explosives and shell fuses for a further seventeen years, that is until 1928. He had become a friend as well as an associate of the Du Pont family, particularly of Pierre S and Francis I du Pont, the latter an inventor in his own right who was pleased to work with Hudson on a number of joint projects, including a 'new theory of radioactivity'. The company was not, however, interested in other of Hudson's inventions, so that in 1912 he presented his torpedo and motorite patents to the US government, the first of several magnanimous gestures which earned him the additional approbation of the authorities.

16. Hiram did not live to see the raids carried out four years later by the German Gotha bombers, twin-engined machines driven by pusher screws and armed with three machine guns, which led thousands of Londoners to seek refuge in the underground railway system.

Hudson and Lilian were now spending as much time as possible at Hopatcong, where they criss-crossed the lake in a smart steam launch and were already leading figures in the community. In 1912 the value of the couple's holdings, including the Hotel Durban, was assessed at around $270,000, and in that year Hudson, as president of the local residents' association, was called to appear before a Senate Committee on Railroads and Canals at Trenton, New Jersey. A particular problem had arisen in the shape of the Morris Canal which, flowing by the lower end of the lake, had been built many years earlier to transport anthracite from the Pennsylvania coal mines to the industrial areas of Newark and New York City. The waterway had since fallen into disuse with the spread of the railways, but the canal company proposed to reconstruct it both for heavy freight and as an auxiliary water supply for the New York conurbation. Since, if approved, the scheme would inevitably involve drawing off the waters of the lake, it was strongly opposed by the residents.

In evidence to the Committee Hudson argued eloquently against the Morris Canal project and for preserving the amenities of the lake as a public park to be enjoyed by residents and visitors alike. He stated that since 1905 he and others had taken much trouble to improve living standards; he personally had spent large sums on road works that were still going forward, and all this would be threatened by any activity which resulted in the lowering of the water level. A few months later, to drive the point home and further frustrate the Morris Canal company, Hudson published a booklet entitled *Lake Hopatcong the Beautiful, a Plea for its Preservation and Dedication as a Public Park and Health Resort for all the People.* This had the required effect, and the campaign was successful in that the Morris Canal scheme was effectively prevented from going ahead, although not until many years later was it finally abandoned.

The year 1913 was an eventful one for the Maxim family circle. In February Hudson and Lilian were surprised when Hudson Day Maxim turned up at their house in Brooklyn. They were also somewhat disconcerted, for Hudson had long lost touch with the boy and his mother, Jennie, whose second marriage was, it appeared, a happy and successful one. Now aged twenty-two and in his final year at Yale, Hudson Day was tall and good-looking, and following in the family tradition he boasted of being the strongest man in the university. On the subject of his academic achievements he was less forthcoming. Quizzed by Hudson, he professed an interest in architecture but admitted that his real aptitude was for athletics and gymnastics. It also became evident that during the coming year he would be looking for a job, and that he had called on his now-famous father in the hope that he might be able to pull some strings on his behalf.

The occasion was an awkward one. Lilian seems not to have taken to the young man, who represented an episode in her husband's past which she thought she had put behind her. As for Hudson, his reaction was one of disappointment: 'I had an idea that he would be a bright, alert, snappy young fellow,' he wrote, 'but . . . he seems to be dull.' Still, having made contact, Hudson Day kept in touch with his father to whom he confided in the summer of 1914 that his ambition was to find some means of breaking into the motion picture industry. As it happens Hudson was at the time a shareholder and director of the Colonial Motion Picture Corporation, to whose board he wrote offering to waive half his salary in return for Hudson Day being taken on 'to learn the motion picture business'. This was duly arranged, although it appears that the young man did not distinguish himself, going on to become a salesman with an indifferent record of success.

In July came the sad news of the death of Dr Durban, which was almost as much of a shock to Hudson as it was to Lilian. For years the Durban family in England had meant more to him than his own family in the States, and he was moved to compose a long effusive poem dedicated to 'William Durban – Father'. Hiram Percy, who remained in close touch with his uncle, wrote from Hartford to commiserate. Some years earlier he had invented a silencer for firearms and formed his own company to manufacture it; he was also interested in radio, operating one of the first amateur stations in America, and in aviation, being a founder member of the Hartford Aero Club. Depressed by the feuding and divisions within the family, Hiram Percy was always on the lookout for opportunities to heal the rifts. One such arose on the occasion of the celebrations planned for the spring of 1914 to mark the centenary of the founding of the township of Sangerville, Maine. Naturally Sir Hiram Maxim, its most distinguished son, was invited, and there was even talk of erecting a statue to him, but he pleaded ill-health and a reluctance to subject himself to the discomforts of the Atlantic crossing.

It is unclear who consulted whom, but the upshot was that Hiram Percy agreed to travel to Sangerville and represent his father at the celebrations. Of the three Maxim inventors he comes over as the most civilised and agreeable, beloved and respected by all who came into contact with him. While his daughter Percy's description is perhaps a little fulsome, there is no reason to doubt its essential truth. Hiram Percy's personality, she wrote, was irresistible:

He had an infectious gaiety and spontaneity; he was a fascinating storyteller; he loved jokes and tomfoolery as long as no one was hurt . . . he was inquiring, full of sentiment, responsive to beauty wherever he found it; unworldly and utterly trusting. He was an inspiration to young people, a marvellous companion, a completely devoted husband and a father never to be forgotten.[17]

After returning from Sangerville Hiram Percy made a determined effort to restore relations with his father, writing a long letter to the old man with a detailed account of the proceedings, which began 'My dear Father, BLOOD IS THICKER THAN WATER', and ended 'Your affectionate son, HPM'. There followed an amicable exchange of correspondence which may well have inspired Hiram to take a more serious interest in the writing of his autobiography. Certainly long distant memories were stirred which in spite of everything that had happened he felt should be passed on to his son: 'Of course,' he wrote to him, 'I should tell you about the kind of life I lived in the town of Sangerville – a poor little bare-headed, bare-footed boy with a pair of drill trousers frayed at the bottom, open at the knees, with a patch on the bottom, running wild but very expert at catching fish . . .'

---

17. *Family Reunion*, p212.

EIGHT

Armageddon and After

*And for all that wastage you should lay the blame
on the Maxim who gave this dreadful gun his name,
Sir Hiram, the ex-Yankee British knight
and his brother Hudson Maxim's Maximite.*
                    TONY HARRISON: 'SQUARE ROUNDS'

One of the few observers accurately to predict the consequences for
the fighting man of advances in modern weaponry before the First
World War was the Polish economist I S Bloch, whose long, rambling
work *The War of the Future* appeared in Paris in 1899. Bloch argued
that magazine rifles and machine guns had become so lethal that
'everybody will be entrenched in the next war. The spade will be as
indispensable to the soldier as his rifle. The first thing that every man
will have to do, if he cares for his life at all, will be to dig a hole in the
ground, and throw up as strong an earthen rampart as he can to shield
him from the hail of bullets which will fill the air'. This together with
the advent of quick-firing guns and high explosive shells meant that
artillery would prove increasingly to be the decisive weapon: '. . . as all
infantry when acting on the defensive will be entrenched, armies in
future will find themselves mainly dependent upon artillery'.[1]

Bloch gives a remarkably perceptive account of the resulting stale-
mate, when wars 'will of necessity partake of the character of siege
operations, each combatant . . . confronted by carefully prepared and
elaborately fortified networks'. His conclusion was that war between
the major powers had become impossible 'except at the price of [econ-
omic] suicide'. Not only were huge conscript armies unmanageable,
but since a war of entrenchments was bound to last for many years the
cost of maintaining them in the field would be such that in the end the
outcome would be decided not so much by the fighting as by financial
and industrial considerations. Should the two sides be equally bal-
anced in their ability to sustain the conflict, then victory or defeat
would depend on the will of the civilian population, that is 'the quality

---

1. Quotations from the abridged English version, *Is War now Impossible?*, edited by the influential
   writer and journalist W T Stead.

of toughness or capacity of endurance, of patience under privation, of stubbornness under reverse and disappointment . . . The men at the front will very speedily be brought to a deadlock. Then will come the question as to how long the people at home will be able to keep on providing the men at the front with . . . what they need to carry on the campaign'.

Bloch's book had the effect of persuading Tsar Nicholas of Russia to summon the first Hague Conference, but there is no sign that it exerted much influence on military thinking. On the contrary, in the years before 1914 a European war was thought distinctly possible, and was anticipated and prepared for in endless war games conducted by the various high commands. Nevertheless, when the storm broke it took everyone by surprise. As the young men flocked to the recruiting centres, the bankers and industrialists were dismayed by the opening of Pandora's Box with all its unpredictable consequences. The deterrent had failed to deter, and prominent businessmen consulted in a last minute attempt to limit the damage. Heading a delegation from the City to Downing Street, the aged Lord Rothschild, who perhaps more than any other individual had fostered the growth of the armaments industry, urged on Prime Minister Asquith the need for mediation and restraint. It was too late. The nations were on course for war, and once set in motion the military juggernauts were driven inexorably forward by 'understandings' between the general staffs and the imperatives of mobilisation timetables and long planned offensives.

As the conflagration took hold and both sides sought to get in a first and decisive blow, the belligerent armies were also taken by surprise. From the beginning the course of the fighting was dictated by the deployment of new weaponry and in particular the overwhelming preponderance of the Central Powers in heavy artillery and machine guns. Only gradually were the Allied armies able to adjust to take account of the power of these formidable weapons. In the east the Russian steamroller was halted by the sheer weight of German firepower. On the western front the Germans advanced to within sight of Paris, only to be halted on the Marne having outrun the support of their artillery, while the French suffered close to a million casualties mounting their own frontal attacks in Alsace and Lorraine. During the autumn and winter, as anticipated by Bloch, the combatants settled in to a continuous line of entrenchments which was to be little changed for the next four years.

The influence of Hiram Maxim's gun in bringing about this state of affairs is well known and needs little elaboration here. From an early stage observers noted how the Germans based their tactics on the coordinated fire of machine guns, riflemen being used as auxiliaries to transport and stockpile ammunition. With heavy mountings set on fixed lines, these weapons could in the words of a recent study:

> . . . beat No Man's Land systematically or skim the whole length of the enemy's parapet at will, in any conditions of weather or light. With a total replenishment and firing crew of less than ten men – and heroically even at times just one man – a single belt-fed Maxim could sweep an area up to 2500 yards deep and perhaps 500 yards wide. In favourable terrain it could halt a whole battalion dead in its tracks by its insistent hail of shots – a complete 250-round belt each minute was by no means a difficult rate of fire to sustain.[2]

In consequence, as shown by bitter experience over many months of fighting, all attempts to break through the enemy's defensive line failed, most being repulsed with heavy loss of life. As to the question which was the more decisive weapon, artillery or the machine gun, the fact is that it was the *combination* of the two that proved to be so deadly. While the greater number of casualties was caused by artillery fire, it was the machine gun, ably abetted by the barbed wire entanglement, which pinned the infantry down and presented the artillery with relatively easy targets. By the early months of 1915 it was apparent that the only means of destroying well-emplaced machine guns was to knock them out with artillery, preferably heavy artillery firing high explosive shells rather than field guns using shrapnel. It was also increasingly obvious to the Allied field commanders that under the conditions of trench warfare more machine guns were urgently needed if their hard-pressed troops were to hold their own. With one machine gun seen as equivalent to fifty infantrymen, Lloyd George calculated that the BEF could make up for its lack of numbers by training 200 machine gunners instead of 10,000 riflemen.

But already the armies on every front were suffering from shortages not only of machine guns but of shells, rifles and war material of all kinds, and, having committed their domestic resources to the full, the British and Russian governments turned for help to the industrial might of the United States. Within months of the outbreak of war Du Ponts found themselves inundated with orders, mainly from the Allies, for high explosives such as picric acid and TNT, which last it had hitherto produced only in quite small quantities. At first the directors reacted with caution. The war was not expected to last long, and in the

---

2. Paddy Griffith, *Battle Tactics of the Western Front* (Yale 1994), p38.

past over-production of military powders had resulted in financial loss. As, however, the flow turned into a flood it became evident that the company was being presented with a commercial opportunity on an unprecedented scale. The existing plant at Wilmington, Delaware, was extended to cover the manufacture of all the chemicals and explosives required, and the flood soon became a torrent.

Since the political situation was far from clear, a certain discretion had to be observed. The sympathies of the American public were fairly evenly divided as between the Entente and the Central Powers: most believed that the fighting in Europe was none of their business and few had any wish to become involved. Ostensibly the United States government took a firmly neutral stance, insisting that licences for war contracts be controlled and issued impartially to the many buying agencies, official and unofficial, which were soon jockeying for position in the industrial marketplace. Nevertheless, large orders were increasingly accepted on behalf of the Allies, even if only because the imposition of the British naval blockade made it virtually impossible to trade with the Germans.

From the beginning Hudson Maxim emerged as one of the leading spokesmen for a powerful pressure group which believed that sooner or later the United States was bound to be drawn into the conflict, and that it should look to strengthen its armed forces against this eventuality. Over the years the subject of war and of preparation for war had become for him something of an obsession. In 1912 he had patented a game of skill 'which shall simulate or symbolise war'. Played on a board with 100 squares, this was based on the principles of chess but seems to have been too complicated to appeal to a wider public. Hudson enjoyed trying out the Game of War on his friends, and after the real war broke out he brought it to the attention of the US chess champion Frank J Marshall, who replied commending it as 'most fascinating and, like chess, admirably adapted to the exercise of the highest skill . . . I trust that it may very soon be put upon the market . . .'.

Hudson was indefatigable in support of the preparedness movement, writing articles, giving lectures attacking the pacifist lobby and calling for a larger navy and an army of a million men. In 1915 his very public patriotism, together with his reputation as an explosives expert, led to his being nominated by the Aeronautical Society of the United States to the Civilian Advisory Board (later renamed the Naval Consulting Board) set up by the government to act as a link between American industry and the armed services. Headed by T A Edison and composed of men with technical experience in a variety of fields, the board was given the task of identifying promising inventions for war use. In due course Hudson became chairman of the Committee on Ordnance and Explosives which during the war examined thousands of ideas from all over the world. 'Everything of which the human imagination could conceive,' he wrote, 'was submitted to us, and some of the things were wild and weird.[3] I think it may be taken as a fair average that there are a hundred crank inventors to every real inventor. Not more than one per cent . . . amounted to anything, and only a few of these were of practical use.'

At the same time Hudson continued to work on his own inventions, which after America entered the war came to include a plan for minimising the damage caused by torpedoes to freight and troopships by collecting and dissipating the gases of the explosive blast and venting them up into the atmosphere. This was duly investigated by the authorities, as were also a position indicator for submarines, a depth charge, a firing mechanism for contact mines and a method of exploding torpedoes under the hull of an enemy vessel. Some of Hudson's ideas do appear to have contributed to ongoing researches. Most, however, were not taken up by the navy, which also failed to show interest in his 'multi-gun concealed turret' or a device described as the 'Dewitz-Gibson-Maxim aerial torpedo system' for dropping torpedoes from aircraft. This last arose from a collaboration with the Danish scientist Baron Hrolf von Dewitz, to whose book *War's New Weapons*, published in 1915, Hudson contributed an introductory preface. Apart from underlining the dangers for America of a lack of readiness for war, the book was markedly anti-Allies and anti-British in tone, von Dewitz being obviously an admirer of the Germans and the scale of their military achievements.

In March 1915 Hudson was invited by Hearst's International to sum up the case for rearmament, a task he found so congenial that he was able to complete the manuscript in just one month. The resulting work, *Defenseless America*, was published in June. Its appearance was well timed, coming as it did soon after the sinking of the Cunarder *Lusitania* by a German submarine, an event which involved the loss of over a hundred American passengers and provoked a strong protest from President Wilson. The book seemed indeed to coincide with a turn in the tide of public opinion, and Pierre du Pont, who had contributed his own ideas and suggestions, was concerned to ensure that

---

3. They included a 'catapult bomb-throwing gun', a 'diving suit equipped with a propeller' and a 'centrifugal disk-throwing mechanism, as attached to a motor cycle'. This last its inventor demonstrated to Hudson at Hopatcong by skimming metal disks into the lake.

the weight of the corporation was thrown behind its promotion. The author was to claim that 50,000 copies were sold 'through ordinary channels', while the Du Pont company subscribed $15,000 to enable thousands more to be distributed to 'all regular army and navy officers of the National Guard and to many college professors, public libraries, high-school teachers, men of prominence and important institutions throughout the country'. A further 10,000 copies were mailed to college graduates, and 8000 were placed in guest rooms beside the Bible in leading hotels.

Defenseless America took the form of a polemic for the armaments lobby and against its isolationist and pacifist opponents. International law, argued its author, was useless unless backed by force, and war, whether with Germany, England or Japan, 'is inevitable and imminent . . . It is criminal negligence for a nation not to be prepared . . . and abreast of the times in arms and equipment'. Noting the ease with which an invading force could knock out America's industrial capacity by occupying an area with a radius of 170 miles from New York, Hudson pointed to the folly of relying on a militia. Success on a modern battlefield, he declared, depended on men trained in the use of modern weaponry which it was too late to produce *after* war broke out. In particular he lamented the inadequacies of the US navy in terms of capital ships, ordnance and stocks of ammunition when compared with any of its rivals. As to the criticism that the armament manufacturers had a direct interest in taking America into the war, Hudson affirmed that he was personally acquainted with many in the trade who were 'among the staunchest of peace men . . . they would be no more guilty of promoting war to bring themselves business than a reputable surgeon would be likely to string a cord across the street to trip up pedestrians and break their limbs in order to bring himself business.'

In assessing the influence of his book on American public opinion, Hudson was to make his customary extravagant claims, more especially after it went on to provide the inspiration for a motion picture entitled *The Battle Cry of Peace*, which, helped by the author's contacts in the movie world, was premiered in October 1915 at the Vitagraph Theatre in New York City. Subsequently the picture (no print of which appears to have survived) went on general release, and in conversation with Clifton Johnson Hudson was to assert that the movie was seen by fifty million people: 'Probably no other one thing did so much to get the United States into the war as did my book, with that great picture sweeping the country.'

In another context the preparedness movement was to backfire against Hudson and cause him much trouble. As the conflict in Europe escalated and the warring governments looked to tap all possible sources of war material, he was consulted by two ex-army officers, Lawrence Angel and Edward Beckert, who proposed to set up a factory to make Maxim machine guns. Hudson had always regarded this as a possibility, hitherto frustrated by his brother's intransigence, and his response was positive. In the rapidly changing war situation, he felt, Hiram could hardly object to such an enterprise, and so he offered his collaboration on condition that finance to back the project was forthcoming. Already a front organisation of the Russian government, the Russian-American Asiatic Corporation, had indicated its readiness to place an order for 10,000 guns, and a New York businessman, Robert Sweeny, undertook to back the venture to the tune of $2 million. In August 1915 the Maxim Munitions Corporation was formed with Beckert as general manager. Factory premises were leased at New Haven, Connecticut, and first steps taken to put in hand the manufacture of '10,000 Maxim guns of model 1904, water-cooled, recoil type, mounted on tripod, as adopted and purchased by the United States Government'.

In September Hudson wrote to Pierre du Pont to let him know that he had 'gone into the munitions business', and to reassure him that this had nothing to do with explosives, which 'we hope to be able to purchase from you'. Busy as he was with many other matters, his commitment to the enterprise was strictly limited. Although becoming a director with a large (if valueless) share issue, he agreed only to lend his name, itself a potent asset, to the Corporation together with the right to develop such of his inventions as were no longer assigned to Du Ponts, namely those relating to 'guns, projectiles, aerial torpedoes, periscopes and position indicators'. In fact the new company's interest was confined to machine guns, and at first the signs seemed hopeful. Investigation of the unexpired Maxim-Vickers gun patents indicated that by introducing some minor design changes infringement might be avoided, and as a first step a sample 1904 model Maxim gun was bought for $1900 from the Springfield Armory.

On the strength of assurances from Russian officials, office space was rented at 52 Broadway, machine tools were purchased and personnel recruited, and negotiations went forward with a view to securing orders from other government agencies. But as the months went by it became apparent that Beckert and Angel lacked the expertise necessary to run any kind of factory, let alone an operation so complex as the manufacture of machine guns. More seriously, the money promised by Sweeny failed to materialise, while the suspicion took hold that he and his associates were adventurers on the make and that the whole enterprise was an exercise in stock-jobbing, that is raising money on the stock market by pushing up the price of virtually worthless shares sold to a gullible public. With Wall Street rising to record highs, fortunes stood to be made from the supply of munitions to the warring powers, and the inducements to invest in projects offering the chance of quick profits were persuasive. In October advertisements appeared in the press announcing that since the Maxim Munitions Corporation 'controls valuable patents for machine guns invented by Hiram Maxim' it was about to go into full production; also that a new gun was being tested at New Haven, and that the Russian government had 'pledged itself to take the entire output of the factory indefinitely'.

Since these claims were to say the least dubious, eyebrows were soon being raised, and Hiram Percy wrote to his uncle from Hartford to express anxiety about the rumours circulating with regard to the MMC. Hiram Percy had earlier developed his silencer for firearms 'to meet,' as he put it, 'my personal desire to enjoy target practice without creating a disturbance . . . For nearly two years I sought for some way to check the powder gases from bursting into the air when the bullet left the barrel, [which] causes the objectionable report.' The answer he found was to 'whirl' the gases in a metal tube, so slowing them down and reducing the bang. The Maxim Silencer Company formed in 1908 had, however, met with little success since the device was seen as an aid to criminals and a menace to public safety. A number of states passed laws prohibiting its sale and several foreign countries prohibited its importation. Only after 1914 did the company enjoy something of a revival by diversifying into the production of other war material such as gas-grenade bodies and scabbards for bayonets.

Invited by Hudson to act as consultant to the Maxim Munitions Corporation, Hiram Percy replied urging caution to safeguard the family name: 'You and I are going to lose the most valuable thing we possess if Maxim Munitions becomes a stock-jobbing venture, and we must be very watchful and frank with each other.' Hudson was quick to take this sound advice, writing to Beckert to convey his concerns about the Corporation's finances and to demand that a lawyer be engaged to ensure that its business was being properly conducted. He also pointed out that it was wrong and misleading to seek to inflate share values by using Hiram's name in publicity material:

> I would not have any business dealings with Sir Hiram Maxim or be associated with him . . . He is my worst enemy, and I have the most utter contempt for him . . . At the present time he is looked upon in England as a joke . . . Not only will we not have the cooperation and

support of Sir Hiram, but we must count upon his most determined
enmity . . . he will do everything in his power to injure us.

These remonstrations fell on deaf ears, and there was worse to
come. The Bureau of Ordnance queried the production of machine
guns for the Russian government on the grounds that this contravened
US neutrality: the sole reason they had made the 1904 pattern gun
available to the MMC, they said, was to assist in its manufacture for
the American service. The question of patents also proved more difficult than anticipated. Notwithstanding Hiram's earlier differences
with Colts, Vickers had authorised them to produce the 1904 model
Maxim adopted by the American army and navy, and British government agencies were also placing substantial orders (which in the event
the company had great difficulty in meeting) for the 1912 model
Vickers gun. In September, therefore, Colts wrote to the MMC pointing out that 'we hold the exclusive right to manufacture this gun
within the United States under licence from Vickers Ltd . . . You will
infringe this contract if you manufacture these guns, and we will immediately take steps to protect our rights.'

Undeterred, Beckert and his colleagues authorised new share issues,
keeping up the pretence that the MMC was going from strength to
strength, and Hudson was soon a worried man. Like his brother he
was something of an innocent where financial matters were concerned, and it was evident that he had got himself into an awkward
situation. His only wish was to steer clear of the whole affair, but this
was easier said than done. Under contract as a director and for a time
president to the Corporation, he had entered into obligations that
could not be shrugged off and which for many years were to threaten
him with legal proceedings and even the spectre of bankruptcy. During the early months of 1916 he was increasingly at odds with his
fellow directors, his patience finally snapping when it was announced
that they were considering adopting the so-called Enricht process
which, it was claimed, could convert water into a fuel to replace gasoline. Dismissing Enricht as a 'faker' and his process as 'a chemical and
physical impossibility', Hudson resigned as director and sent a circular
to newspaper editors disclaiming any further connection with, or responsibility for, the doings of the MMC.

But much damage had already been done, not least to Hudson's
endeavours in the cause of national preparedness. In the spring of
1916 he collected material for a Handbook entitled, somewhat uncon-
vincingly, 'Leading Opinions both for and against National Defense, a
symposium of opinions of eminent leaders of American thought . . .

presenting both sides of the question with absolute impartiality'. His own contribution consisted of a vigorous attack on an alarming proposal being put forward by the pacifist lobby to 'take the profit out of war' by following the British example and bringing the armaments industry under a measure of state control. Describing any such move as 'colossal folly', he advised that the US government should rather emulate the Germans, whose war effort depended on cooperating with and encouraging great industrial concerns such as Krupps.

The reaction of Hudson's opponents to all this was predictable enough, and they became ever more outspoken in questioning his motives, alleging that his real objective was to swell the profits of the armaments industry. In April Henry Ford took a full-page advertisement in the *New York World* to launch a concerted campaign against Hudson's much publicised lectures and also against his *Defenseless America* and the moving picture based upon it. In his reply ('A Message to Patriotic Americans') Hudson accused his detractors of slander and even treason. He made no apology for having gone into the business of arms manufacture, and insisted that he had nothing to hide. But his critics were in no way convinced, continuing to maintain that Hudson's initiatives were thinly disguised exercises in self-interest, aimed at boosting the sale of armaments and in particular the stock of Du Pont and the Maxim Munitions Corporation.

Some went further, questioning the propriety of retaining Hudson on the Naval Consulting Board. Fortunately most members of the Board were sympathetic to Hudson's point of view, and he could count on the support of several powerful colleagues, among them Edison and Secretary of the Navy Josephus Daniels. Partly perhaps because he was anathema to Hiram, Edison was admired and cultivated by Hudson, who was accustomed to praise his achievements in extravagant terms: 'No other man has so profoundly influenced the industrial development of his time . . . If a Joshua could command the inventions of Edison to stand still for an hour, the entire industrial activity of the country would be held up and the effect would be quite as cataclysmic as the holding up of the sun in its course.' Edison responded in kind, describing Hudson as 'the most versatile man in America'. He and his colleagues were well aware of Hudson's record as a research scientist and of the contribution he had made to explosives technology. They were also appreciative of his work for the Committee on Ordnance and Explosives and of his ongoing researches which were throwing out a stream of original, if not always practicable, ideas.

Throughout the war and for the rest of his life Hudson continued to draw $500 a month as consultant to Du Ponts, which valued him as much for his public-relations initiatives as for his achievements as an inventor. In 1913 the company's gross receipts amounted to $27 million; in 1915 they had risen to $131 million and in 1916 to $319 million. After the United States entered the war in April 1917 they rose still higher, and during the whole period of the conflict the company was the largest single supplier of propellant and high explosives to the Allies. For this massive effort Hudson was to claim a due share of the credit, which was true in the limited sense that he had contributed to the company's research programme, and continued to profit indirectly from its vastly increased sales. Over the years the original Maxim-Schupphaus powder had undergone modifications with a view to making it safer, as also had the high explosive shell fillings based on his maximite, which before 1914 was gradually replaced by the less volatile TNT and ammonium picrate.

As the war progressed, therefore, Hudson had every reason to believe that he had succeeded in keeping pace with Hiram, which was always one of the objects of his life. Although machine guns deriving from the old man's brainchild were dealing out death and injury on a previously unimagined scale, Hiram had ceased to be other than an occasionally active spectator. During 1914 he and Sarah were mainly occupied with the writing of his autobiography, which was published by Methuen in the following year. It was received with respect, although in the light of the carnage on the western front and at Gallipoli, reviewers thought it advisable to emphasise the author's wider achievements rather than the gun. Comparing Hiram to Conan Doyle's quixotic Brigadier Gerard, at once outrageous in his conceit and admirable in his style, the *Times Literary Supplement* went on to comment:

> One is carried away by the bravura of his methods to the point of forgetting his solid qualities; but the foundation of his success is that of the industrious apprentice. [Maxim] possessed great personal strength, a willingness to work sixteen hours a day, and a capacity for assimilating what books had to teach him. He thus accumulated knowledge both practical and theoretical and, with this to draw upon, his keen brain found its way to the heart of the problems he was asked to solve.

Not that Hiram was by any means an extinct volcano. As ready as ever to rise to fresh challenges, in November 1914 he volunteered his services to Lord Moulton, newly appointed as Director of Explosives Supply, who, while tactfully declining the offer, commented that the inventor 'was still the youthful and eager optimist that I had always found him'. In May 1915, shocked by the German use of poison gas at the second battle of Ypres, he lodged a provisional patent for an 'apparatus for dissipating noxious gases', which worked by creating 'large and rapidly spreading fires . . . in the path of the advancing gas, causing an upward rush to drive [it] up and out of harm's way'. This was tested by the experts at the War Office and rejected by them although it was later noticed that the Germans were lighting fires along the parapets of their trenches to make the gas cloud rise and disperse. Another proposal put forward by Hiram was for using incendiary bullets against Zeppelins: 'Each cartridge would carry a trail of fire containing pulverised magnesium which would be sure to ignite the hydrogen if a hit were made.' A further stratagem was to 'fly over the Zeppelin and let down a bomb attached to a cord or wire, the bomb being provided with a lot of sharp hooks . . . by dragging this across the Zeppelin some of the hooks would catch . . . causing the bomb to explode.'[4]

Hiram was now in his seventy-sixth year. While Hudson was consolidating his status as a celebrity his brother had been for all practical purposes sidelined, and soon he was to make his final exit. On 1 July 1916 the British and French armies launched their long prepared 'Big Push' at the Somme, which was intended to deal a decisive blow, breaking the German resistance and bringing the war to an end. The terrible casualties suffered by Kitchener's armies on the first day of the battle and the sad history of the long drawn-out campaign that followed have entered into the public consciousness as epitomising the futile carnage of the Great War. Much of the slaughter was caused by the failure of the preliminary artillery barrage either to cut the German wire or to knock out the concrete bunkers and carefully constructed dugouts at the heart of the enemy's defensive position. Consequently the German machine guns, skilfully sited in depth, were able to do their work with clinical efficiency, and the infantry were scythed down like so much corn as they advanced in extended lines across No Man's Land.

No record has come down to us of Hiram's reaction to these grim events. Like his brother Hudson and so many other of his colleagues in the armament business, he had always argued that his inventions would make war impossible. Now that he had been proved so spectacularly wrong, there was nothing he could do about it except trust

---

4. This proved impractical for the simple reason that at the time the Royal Flying Corps possessed no aircraft capable of outclimbing a Zeppelin.

that his inventions, and especially his machine gun, would in the end help the right side to win. This they were in due course to do, as hundreds of factories in Britain and America turned out shells and explosives, and thousands of Vickers and Maxim guns were produced at Erith and Crayford as well as by the Royal Small Arms factory at Enfield and the Colt company in the United States.

Hiram himself was not to live to see the massive increase in manufacturing output which was in the end to achieve victory. With the outbreak of war he and Sarah had sold Thurlow Lodge and moved to a smaller house, Sandhurst Lodge in Streatham High Road, where in the spring of 1916, even as the great battles raged in France, the inventor dictated the notes on the basis of which Mottelay was to produce his biography four years later. Even in old age, it would seem, he could not resist indulging his addiction to practical jokes. According to a rumour which reached the press, the purchasers of Thurlow Lodge were alarmed to find a note from Hiram informing them that he had some years earlier buried a trunk containing nitroglycerine in the grounds. Although he could not recall the exact spot, he wrote, it was important that every effort be made to locate and remove the trunk and with it the danger of an explosion. The unfortunate new owners were therefore obliged to dig up several acres of lawn and gardens, but no trunk was ever found.

At the same time Hiram was absorbed in a new project designed to remedy the growing shortage of fuel for lorries and other military vehicles by finding a means of converting common paraffin into cheap and serviceable petrol. To this end he rented the old Sorting Post Office at Herne Hill, which he adapted as a laboratory and equipped with the aid of the ever supportive Albert Vickers. In accordance with his normal custom, Hiram took the precaution of securing provisional patents ('Improvements relating to the production of light mineral oils . . . and to the conversion of heavy hydrocarbons into lighter hydrocarbons'), but then with the onset of autumn he was once again assailed by his old enemy bronchitis. This was soon complicated by pneumonia, and, despite all Sarah's ministrations, the combined attack proved fatal. Early in November the Somme battle petered out in the mud, the opposing armies having fought themselves to a standstill, and on the 24th the old man died at his Streatham home.

The coincidence did not pass unnoticed in the rash of obituaries that followed. 'Maxim,' commented *The Engineer*, 'was of a type which only the United States has so far succeeded in producing . . . a man . . . of vast tenacity of purpose, of infinite self-reliance and of unbounded self-esteem.' Notwithstanding his many and varied inventions, however, history would be bound to endow him 'with the unenviable reputation of having been directly and indirectly responsible for depriving more men of their lives . . . than any other one man.' Others were prepared to take a broader view. *The Times* referred to Hiram as 'perhaps the most accomplished mechanician of his day, [who] extended his inventive faculties over an extraordinarily wide range'. And Lord Moulton, in a better position than most to offer an opinion, was later to sum up the significance of his achievement with legal precision: 'In several of the great movements of his time he took a large share, and in all of them he left his mark'.[5]

As he had foreseen, and in accordance with his wishes, Hiram was buried in the large cemetery at West Norwood. The funeral was a quiet one, attended, apart from Lady Maxim and the fourteen-year-old Maxim Joubert, only by Lord Moulton and one or two close friends and representatives of the Vickers company. In his will, published in December, he left the bulk of his gross estate, valued at £33,000, to Sarah or in trust for Maxim Joubert. In addition legacies of £1000 each were given to Mrs Josephine Lewis and to Mrs Romaine Dennison of Mount Vernon in the state of New York. In May 1909 it had been mentioned in *The Times* Court Circular that Sir Hiram, travelling alone, had just returned from New York on the steamer *Minnehaha*. The reason for his visit at this time is unclear, and it is tempting to speculate that it was to attend the wedding of his daughter Romaine to Mr Dennison, after which the couple settled down to live in the United States. As for Mrs Lewis, her name is not familiar from any earlier record, but the assumption in the family has always been that she, too, was Hiram's natural daughter.

After the war Hudson and Lilian looked forward to a long and agreeable retirement. In 1919 they sold their house in Brooklyn (they had been trying for many years to dispose of it and now did so at a loss, 'for the servant question is such that it makes living in a large house very troublesome') in order to base themselves permanently at Lake Hopatcong. As president of the Inter-Municipal Park Commission Hudson continued to be busy with local affairs, while he and Lilian together looked after the extensive properties around the lake. They were regarded by all as the very model of a devoted couple. Lilian has been quoted as saying that she would 'rather be Hudson Maxim's slipper bearer than Queen of England', and every day that Hudson was away he wrote her long sprawling letters starting 'My Dear Blessed Sweet Lill' and ending 'All love always . . .' In 1921 Hiram

---

5. Mottelay, op cit, Introduction, xvii.

Percy's daughter Percy, then aged fifteen, visited Hopatcong with her parents. Hudson and Lilian, she recalled, 'put on their extra best manners for my mother [Josephine]. Hudson was a most impressive man . . . very handsome, just under six feet, with a big frame and . . . exceedingly strong and husky.'

Constantly receptive to new ideas, Hudson decided early in 1918 to capitalise on what was a long-standing dedication to food and healthy cooking. In particular he believed that the humble soya bean, recently imported into America from China, was being neglected as a source of protein at a time when droughts in the mid-western states were decimating traditional grain harvests. Experimenting in his kitchen at Maxim Park, he came up with two soya-based products, 'Maxicream', designed to rival peanut butter as a sandwich filling, and 'Maxim-Feast', a multipurpose complete food suitable for canning which he sought to market as a source of nutrition to starving populations around the globe. His first thought was that the army might wish to lay in supplies for feeding the troops, a notion which he put to the authorities but with the ending of hostilities was politely declined. Nor did any of the big corporations take up his soya products, which only recently have become established in the form of 'tofu' and as a principal constituent of processed foods.

Hudson had now modified his views on the need for armaments, writing that 'I have not changed from militarism to pacifism, but the World War has altered circumstances, and the supreme need is now for an armament truce and not for more armaments', which simply put up taxes. He continued, however, to be actively involved with the Du Pont company on research into such developments as a new flashless propellant, and to benefit from the prestige this conferred on him. Thus in 1919 he was called by the Lehigh Valley railroad company as an expert witness during an investigation into the massive 'Black Tom' explosion which took place in New York harbour in July 1916. This had resulted in wholesale destruction of barges and dockside warehouses, where a cargo of TNT destined for the Allies was in process of being loaded on to ships. Damage had been caused over a wide area, blowing out windows in the skycrapers of New York City, and the consequences from an insurance point of view were serious. The issue was whether the railway company had caused the explosion by careless handling or failure to deal with an outbreak of fire, or whether the disaster had followed an act of sabotage by German agents.

On behalf of the railroad company Hudson stated that TNT, being essentially inert, was most unlikely to blow up as the result of a fire or of careless handling. It is, he declared, 'very doubtful if a million pounds of TNT piled into one great mass and ignited would detonate. It would almost certainly burn up without detonation unless detonated by some external exploding means.' His evidence helped to exonerate the company from blame, and also to confirm suspicions about the activities of the German embassy and its agents during the war. There is little doubt that determined and on occasion successful attempts were made to prevent war material reaching the Allies, many involving the planting of explosive devices in the hold of ships. These were extremely difficult to detect since, being timed, they usually went off in mid-Atlantic and so could easily be mistaken for the strike of a mine or torpedo.

In 1921 the continuing interest in Hudson's wartime inventions was reflected in an agreement signed by the Acting Secretary of the Navy, enabling the authorities 'to make and conduct such experiments and tests as [they] shall deem necessary to demonstrate whether or not the said inventions possess sufficient value and usefulness to warrant their adoption.' During these experiments and tests Hudson was to be consulted and his expenses paid, even though 'he does not agree to devote his entire time and attention to the said work'. The proviso indicates a changing attitude on the inventor's part, away from military hardware and, under the influence of Lilian and Hiram Percy, towards the cultivation of a more philosophical, artistic and cultured way of life. He had already developed an interest in psychology, contributing a pamphlet (on the *Practical Psychology of Cooperative Conduct*) to a series published in 1920 by the Society for Applied Psychology, and was seeking further to widen his intellectual horizons.

For some time Lilian had been urging Hudson to write his autobiography, partly because she thought he had much of interest to say, partly to correct the distortions contained in Hiram's book and set the record straight. In fact he had been keeping notes on his own and the family's history since before the war, and these he was able to talk over with Hiram Percy on the occasion of the latter's visits to Hopatcong to play tennis and discuss the state of the world. But as is often the case, when he tried to order his thoughts into a connected narrative, his natural ebullience refused to transfer itself to the printed page. In 1923 he contacted Clifton Johnson of New York to write the book for him on the basis of interviews at Hopatcong and in the course of a motoring tour of Hudson's old haunts in Maine. In this way, as Johnson stated in his preface, Hudson's life story as well as samples of his 'spiced and peppered' conversation and 'irresistible humour' could be recorded for posterity by a stenographer and by Mrs Maxim, 'whose expert and sympathetic aid in preparing the manuscript has been invaluable'.

Responding to Hudson's invitation, Johnson travelled by rail to Landing, the stop for Lake Hopatcong, one afternoon in June:

> When I left the train, I found myself at the southern tip of the lake and learned that I could go to the Maxim home, three miles distant, by a passenger motor boat. I stepped on board, and in a short time the boat approached a little cove that lay just back from a rocky promontory; and there, well up a steep slope dotted with great trees, stood Mr Maxim's house. It was big, substantial, castle-like in architecture and setting, evidently built in instalments, and rather interestingly incongruous.

Inside he found himself in a large dining room 'with a gloomy oak-timbered ceiling and an enormous fireplace', and around the dining table 'sat a group of persons of various ages, the central figure among whom was a man of burly form with a shock of curly white hair. Surely he was a medieval baron and these others were his family and retainers. One of them was the beautiful lady of the castle, Mrs Maxim, and the baron was Hudson Maxim.'

Clearly Johnson was impressed, even intimidated, by the force of Hudson's personality, which dominated the household as it did the local community. The sessions at Hopatcong never lasted long for his host's 'abounding vitality set such a strenuous pace . . . that at the end of a few days I was ready to withdraw to recuperate'. Partly for this reason the resulting memoir is weakened by its anecdotal nature and by its author's willingness to accept all his subject's pronouncements at their face value. Nor did he make any attempt to evaluate Hudson's inventions, preferring simply to draw attention to their importance as reflected in the fact that 'he received for them from the Du Pont Powder Company a lump sum of two hundred and five thousand dollars, besides lesser payments, and a salary as consulting engineer that still continues'.

Nonetheless, the book is revealing about Hudson's lifestyle and his views in general, some of which were ahead of their time, others merely idiosyncratic. Guests arriving at Maxim Park were disconcerted to find notices in their room forbidding the use of perfumes and cosmetics, which, their host declared, were toxic and apt to give him severe headaches. Nor were visitors left in any doubt about Hudson's aversion to cigarette smoking. During the war, he averred, much injury had been inflicted on the men at the front by supplying them with free cigarettes: 'The permanent effects of cigarette poison are worse than those of poison gas . . . The yellow finger-stain is an emblem of deeper degradation and enslavement than the ball and chain'.

At the same time Hudson's reported conversation reflects a thoughtful outlook based on sound common sense, even a certain home-spun wisdom. 'Happiness,' he declared, 'consists in the pleasurable exercise of faculty. We like best that which we are best fitted to do.' As for our main asset in life, this 'consists in the goodwill and companionship of those about us'. And no doubt from heart-felt experience he suggested that: 'Genius needs obstruction. Then it goes to work and does things . . . Our great geniuses have owed their greatness mainly to the difficulties that have opposed their progress.' On the other hand, 'I am not a believer in the unqualified advantages of early poverty. I saw and felt too much of it. My dreams are still frequently haunted by its horrors, and when I awake and find myself a rich man, I am enormously relieved. What I have accomplished has been in spite of poverty, instead of by its aid.'

On the question of religion, which he defined as 'constructive good conduct', Hudson acknowledged that religious people often showed 'a fine sympathy' but it was, he believed, 'their goodness that makes them religious rather than their religion that makes them good'. As to the prospects for scientific progress, he was broadly optimistic. 'No man,' he told Johnson, 'is able to foretell the future except from his knowledge of the present; and our knowledge is so small and our powers are so finite that we can only speculate and generalise. Yet it is safe to predict that man's advancement from now on will be more rapid than it has ever been before.' This would probably be helped by new sources of power, for example 'the internal molecular energy of matter [which] is inconceivable in amount . . . if ever we succeed in harnessing it, we shall be able to light, heat and run the world from the dynamo . . . It is estimated that in a single gram of lead there is atomic energy enough to equal the heat which would be evolved by the combustion of three and a half million tons of coal.'

Despite its obvious shortcomings, *Reminiscences and Comments* succeeds in conveying a lively impression of Hudson and helps to explain why he was held in such high esteem. Noting that he was recognised everywhere he went, Johnson quotes with approval the verdict of a psychologist who said of him:

I never before met a man who combines so perfectly the essentials of a scientific intelligence, and those elemental qualities of strength, candour and directness of character that are almost as primitive as the forces of nature. He is a veritable 'cave man' who has stepped forth from an age when the world was young and unconventionalised.

On its publication in 1924 the book received generally complimentary reviews, although Lilian had to restrain Hudson from firing off vituperative rejoinders to pacifists and others who ventured to criticise it, many of his drafts being annotated with a scribbled 'Did not get by censor'. She also kept cuttings of the more favourable comments, including one by a librarian who wrote that it was 'the record of a real self-made man, and one who will join that small circle of the elite, composed of Edison, Ford, Firestone and Roosevelt, first in the affections of the American people.'

As is evident from Johnson's account, Hudson and Lilian contrived during these post-war years to lead fulfilled and happy lives at Lake Hopatcong. From 1921 Hudson appeared as King Neptune in the annual Atlantic City pageant, but in 1924 he declined a third invitation because of the heat and the crowds and the stifling costume he had to wear. As an establishment figure, a friend of artists and writers, a staunch Republican and member of the Moose Lodge and the Sons of the American Revolution, Hudson had a wide circle of correspondents. He and Lilian were also heavily engaged with looking after their property, their main concern being to safeguard the quality (and therefore the value) of residential development by resisting the subdivision of building plots and restricting the encroachment of industry and commerce around the lake.

There were, however, problems, not least Hudson's reluctant but continuing involvement in the affairs of the ill-fated Maxim Munitions Corporation. This was a constant source of anxiety. Having struggled on during the war and even succeeded in producing some explosives for export, the company was in 1919 renamed the Maxim Corporation only to be subsequently declared bankrupt. Thereafter Hudson's finances were regularly threatened by creditors pressing for compensation, claims which he steadfastly resisted since to give in would be to accept a liability which he refused to admit. The Corporation, he insisted to his attorney, had 'utterly failed to provide the money agreed upon, and tried to raise the money by selling stock on the curb. I left the Corporation because of their bad faith with me, and disposed of my stock for a trifle.' His hands were, therefore, clean and he was not prepared to compromise.

In the autumn of 1924 Lilian wrote to her sisters in England to report that following a bout of strenuous activity at Lake Hopatcong supervising the clearing of land for road works, Hudson had become ill with 'indigestion, shortness of breath and anaemia'. He therefore consulted a 'noted' diagnostician who examined and X-rayed him before recommending that he 'eat heartily of nutritious food, especially red beef, and take a preparation of arsenic'. This regimen proved so helpful that he gradually regained his strength, and the following spring he was able to write to a friend: 'I am now 72 years of age and running a footrace with the Great Reaper. We had it neck and neck for a while last winter but I have, with the aid of a good doctor, left the old rapscallion considerably in the rear.'

Nevertheless, Hudson decided to reduce his workload, advertising 500 acres of building land around the lake 'at exceptionally low Bargain Prices', and to spend more time writing and relaxing with friends. These included such luminaries as the sculptor Emil Fuchs, who modelled Hudson's head and hand, and the artist William Oberhardt, who had illustrated *The Science of Poetry* and now painted his portrait. By the summer of 1925 Hudson was writing that he and Lilian were 'as full of engagements and business and bustle as a coconut is full of meat', and he even found time to reconsider his design for a torpedo-proof ship, in which 'pulverised coal alone was contemplated as a buffer cargo to arrest the blast of the torpedo'. He also produced an article entitled *Science probes the Future* in which he anticipated not only atomic power but the coming of super-highways, worldwide air travel, television (that is, 'speaking, stereoscopic motion pictures, in natural color, transmitted by radio'), and a second world war within fifty years. It was a bravura performance, indicating that his mind was as sharp as ever, and during that autumn the *Boston Sunday Post* ran a series of respectful features on the man and his work.

Hudson's health, however, continued to cause concern, and so he and Lilian decided to travel to Europe for a complete rest, at the same time seeking a second medical opinion and avoiding further legal proceedings being taken by stockholders of the Maxim Corporation. In December 1925 they made the Atlantic crossing on the liner *Berengaria* with the intention of spending a few weeks in London and then going on to Paris. It was a time for mending fences, and Lilian wrote home to a friend: 'A week ago we visited Lady Maxim. Mr Maxim thought we ought to drop her a line saying we were in London . . . Immediately on receipt of our letter she phoned and asked if we would go out to Streatham to tea, which we did. She seemed delighted to see us, so that family hatchet is buried.' Hudson also wrote to inform Hiram Percy, who was about to head an American delegation to the first congress of the International Amateur Radio Union in Paris, that they had called on Sarah more than once, so 'everything is all very nice and very pleasant in that quarter'.

In London Hudson was reassured by the doctors about the state of his health, and he also took the opportunity to have his leather hand

replaced by 'Best-Form' of the Aldwych, specialists in light steel artificial limbs. Staying at the Great Russell Hotel in Bloomsbury, he and
Lilian indulged themselves by drinking large quantities of alcohol:
Hudson had always been an active opponent of prohibition in America
and now, judging from the hefty bills for wines and spirits which they
retained as a souvenir, they were more than pleased to make up for lost
time. Just the same Hudson continued to feel 'rotten' and so they
cancelled their Paris trip, returning in February to Hopatcong. Here a
letter was waiting for Hudson from a woman neighbour of his brother
Sam, who, finding him destitute, had got together a few dollars to buy
him some clothes for Christmas. 'You cannot and do not realise,' she
wrote accusingly, 'the pitiful, poverty-stricken plight of Sam Maxim,
your brother!!'

The plea struck home. Hudson had indeed virtually ignored his
once-favourite brother since the family quarrels of twenty-five years
earlier, and now, pressed by Lilian, he arranged for him to be paid a
modest pension of $25 a month. Although later supplemented by Lady
Maxim, this sum was barely enough to sustain the unfortunate Sam,
whose eyesight was failing, and who wrote asking whether Hudson or
Hiram Percy might be willing to purchase the homestead at Wayne.
But in his reply Hudson was dismissive: 'There are a thousand and one
reasons why we should not care to own the place. Possibly Hiram
Percy might be glad to avail himself of the opportunity which you
offer.'

In fact Sam was to outlive his brothers as well as his sisters. Mentally, Hudson was as active as ever, attending dinners, delivering
speeches and dictating articles to a succession of secretaries. In 1925
he spoke at the dedication of a new dam regulating the flow of water
from Lake Hopatcong, and in the same year he returned to his earlier
preoccupation with atomic theory, producing a piece on 'The Ultimate Nature of Matter and Principle of Force, including the Funda-
mental Philosophy of Relativity' and claiming, with the support of the
Scientific American Monthly, that his article of 1889 had gone a long
way to anticipating Einstein. And in 1926 he summed up his talks and
writings opposing the Volstead Act in a pamphlet with the arresting
title 'Some Thoughts and Talks on Prohibition: Freedom Strikes
when Representative Government Falls'.

Hudson's physical condition, however, was steadily deteriorating
under the influence of what his doctors continued to diagnose as
anaemia, but which Lilian began to suspect was something more sinis-
ter. Irrepressible to the end, he dictated a last testament on the subject
of religion entitled 'Looking into Death' which in accordance with his

instructions was published after his death in the magazine *Plain Talk*. As he grew more and more tetchy and out of sorts, Lilian proposed that the two of them undertake an extended automobile tour of the west coast. Her expectation was that the sunshine and clear air of California, together with a health-enhancing diet, would help restore her beloved husband to his normal robust self, but in January 1927, mid-way through their itinerary, he took a turn for the worse, and they had no alternative but to return home. By now it was all too apparent that Lilian was right and the doctors wrong, and that Hudson was suffering from cancer. Very soon he had to take to his bed, and on 6 May he died.

News of his death aroused widespread public concern, and Lilian was overwhelmed with letters of condolence. Doubleday, Page hastened to put in hand a second, cheap edition of Clifton Johnson's book, and from Du Ponts the vice-president, H Fletcher Brown, sent his commiserations: 'It is with the greatest regret that we have learned of the death of Mr Hudson Maxim, who was for so many years connected with this Company and was held by us in high regard . . . we enclose herewith the usual salary check for the present month in the full amount.' Hudson was buried at Maxim Park, his grave marked by a commemorative plaque which was later removed to the site of the dam by the abandoned Morris Canal, where it can still be seen. After the funeral Lilian wrote to a friend:

> I have had thirty-one years of ideally happy married life, with a husband who so adored me that he could hardly bear to have me out of his sight – and this is more than falls to the lot of most women. I have nothing but beautiful memories of our married life, and I am trying hard to do as he would want me to do now – take up the threads of life and make the best of things, work hard, and finish up certain matters he had started. I am not even wearing mourning, because he hated it so.

Lilian and Sarah had always got on well enough despite the animosity between their husbands, and although Lady Maxim, for reasons that are unclear, went on refusing to have any dealings with Hiram Percy. After the two women met in London two years earlier they had continued to correspond, with Sarah, curiously, still heading her letters 'Dear Ique and Lillie'. Now, doubtless with a certain sense of fellow feeling, Sarah wrote implying that Lilian had done well to cope with a wayward husband who had had to surmount so many difficulties in his life. In her reply Lilian rejected any such notion with characteristic spirit: 'I desire no credit for my service to him. It was a joy. He was a great lover, a true and steadfast friend, a hard but square fighter when opposed . . .' Without Hudson, however, she found it difficult to deal with the continuing demands of the Maxim Corporation and the considerable burden of administering the estate at Lake Hopatcong, which included 600 acres of land as well as the Hotel Durban and four cottages. In November she went to England to stay with her sister Pollie in Catford, and on her return she placed the Hopatcong estate in the hands of a New York law firm.

Within a few months Lilian, in her fifties still an attractive woman, had become friendly with a partner in the firm, Michael Dee, whom she married in 1928 and with whom she settled down to start a new life. Apart from the family house, the properties at Hopatcong were in due course sold off and the land sub-divided into ever-smaller plots, so that now the lakeside is crowded with holiday homes and the visitor has difficulty in getting down to the water's edge. Hudson would not have approved, especially as Maxim Park has, apart from a few stone walls, quite disappeared together with most of the original large houses. Otherwise he and Lilian are commemorated only by the plaque and a few place names, including Hudson Avenue, Maxim Drive, Durban Road and Dupont Avenue. It is also pleasant to record that Hudson is today remembered as a local benefactor, not least for his long campaign against the Morris Canal project, and that the community school still bears his name.

What, then, to make of the long and bitter rivalry between the two men? How far was Hiram justified in believing that Hudson had profited from pirating his inventions, and how far was the latter justified in maintaining that he had, virtually unaided, carved out a separate career for himself in America? It is not an easy question to answer. There seems little doubt, however, that while only Hudson benefited from any kind of formal education, this was hardly sufficient to account for the remarkable proficiency in chemistry and engineering which during the 1890s enabled him to succeed on his own terms, and this despite the loss of his hand. Among the Hudson Maxim Papers is a much-thumbed bound volume of Hiram's patents which shows every sign of having been studied with the closest attention, and suggests that they acted in effect as blueprints, inspiring initiatives which the younger man was shrewd enough to develop with the help of able collaborators such as Mowbray, Schupphaus, Alger and Leavitt. The conclusion must therefore be that Hiram was the acknowledged master: Hudson was always the disciple, the pupil quick to learn, but he lacked the instinctive gifts which established his brother as the only true and original genius.

The rest is soon told. Albert Vickers stood down as chairman in 1918, leaving it to the next generation of the family to manage the fluctuating fortunes of the Vickers company during the inter-war years. Hiram's friend and associate Basil Zaharoff, having made a considerable fortune from the arms trade in the decade leading up to 1914, became during the war an intelligence agent for the Allied cause in the Middle East, for which services he was knighted by Lloyd George. He died in the south of France, covered in honours, in 1936.

As for the Maxim family, Sam died in 1928 and was buried near his parents in the cemetery at Wayne. Hudson Day Maxim married and had two daughters, otherwise making no great mark in the world. Lilian Maxim's second marriage lasted to her death in 1953, when the homestead at Lake Hopatcong was sold and its contents auctioned off, two early Maxim guns being snapped up by a collector for $150. Lilian's brother Will tried to make a career in the English Potteries, but he remained a sad case, writing from Hanley in one of the last extant letters to his sister: 'God only knows what a lonely soul I am!'. Hiram's granddaughter Percy Maxim Lee, the only direct descendant of the brothers, is happily still with us and living in a house full of Maxim memorabilia at Mystic, Connecticut.

Lady Sarah Haynes Maxim, perhaps the most enigmatic figure in this long history, lived on with her memories into her eighties and another world war, steadfastly declining to give interviews or to divulge anything of interest about her husband. When she died in 1941 she was buried beside Hiram in the cemetery at West Norwood, as many years later was Colonel Maxim Joubert, who in this way underlined the affection and regard he clearly felt towards his adoptive grandparents. After Sarah's death a London newspaper described her as 'a woman of unusual social gifts which were of great aid to her husband in his career, especially in England, as also was her ability, most unusual in those days, to write short-hand.' However, when in 1926 the twenty-year-old Percy visited London and called to see her, she formed a rather different impression. Lady Maxim gave her tea and presented her with a large diamond pendant, but the house she was living in 'seemed to me incredibly dark and gloomy and cold; she presented a formidable and angular figure – all closed in. There seemed to be no warmth, and it was impossible to imagine her as a person of "unusual social gifts" '.[6]

6. *Family Reunion*, p165.

Herein, perhaps, lies one reason why in the end Hudson turned out to be the more fulfilled of the two brothers. Certainly he counted himself the more fortunate. 'As fate would have it,' he told Clifton Johnson in an echo of the old rivalry, 'I got a better education than Hiram, I have made more money than he ever made, and possibly I have equalled him in reputation.' Ironically it was Hiram who, despite all his achievements, seems to have been less contented, probably because his expectations, being that much higher, were more easily disappointed. 'When I was a young fellow,' Hudson wrote to a friend towards the end of his life, 'it was not my . . . principal aim to accumulate large wealth or to become noted as a rich man. My chief aim was – my biggest ambition was – to become a leader in art, science, discovery, invention and literature, and I have succeeded in each of these endeavours.' He could not have done so had it not been for the Durban family and especially the influence of Lilian, who softened his rough edges and broadened his outlook and gave him an inner peace denied to the older, and greater, man.

The Maxim Family

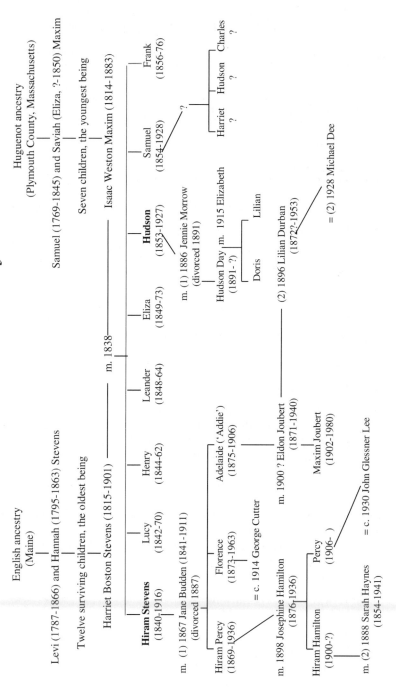

English ancestry
(Maine)

Huguenot ancestry
(Plymouth County, Massachusetts)

Levi (1787-1866) and Hannah (1795-1863) Stevens

Samuel (1769-1845) and Saviah (Eliza, ?-1850) Maxim

Twelve surviving children, the oldest being

Seven children, the youngest being

Harriet Boston Stevens (1815-1901) ——————— m. 1838 ——————— Isaac Weston Maxim (1814-1883)

Hiram Stevens
(1840-1916)

Lucy
(1842-70)

Henry
(1844-62)

Leander
(1848-64)

Eliza
(1849-73)

Hudson
(1853-1927)

Samuel
(1854-1928)

Frank
(1856-76)

m. (1) 1867 Jane Budden (1841-1911)
(divorced 1887)

Harriet   Hudson   Charles
?   ?   ?   ?

m. (1) 1886 Jennie Morrow
(divorced 1891)

m. 1915 Elizabeth

Hudson Day
(1891- ?)

m. (2) 1896 Lilian Durban
(1872?-1953)

= (2) 1928 Michael Dee

Doris   Lilian

Florence
(1873-1963)

Adelaide ('Addie')
(1875-1906)

= c. 1914 George Cutter

m. 1900 ? Eldon Joubert
(1871-1940)

Hiram Percy
(1869-1936)

Maxim Joubert
(1902-1980)

m. 1898 Josephine Hamilton
(1876-1936)

Hiram Hamilton
(1900-?)

Percy
(1906-)

= c. 1930 John Glessner Lee

m. (2) 1888 Sarah Haynes
(1854-1941)

Select Bibliography

Works by Sir Hiram S Maxim
Monte Carlo Facts and Fallacies (London 1904)
Artificial and Natural Flight (London and New York 1908)
A New System for Preventing Collisions at Sea (London 1912)
(ed), *Li Hung Chang's Scrapbook* (London 1913)
My Life (London 1915)

Works by Hudson Maxim
The Maxim Aerial Torpedo: a new system of throwing high explosive from ordnance (London 1897)
The Science of Poetry and the Philosophy of Language (New York and London 1910)
Defenseless America (New York 1915)
Dynamite Stories, and some interesting facts about explosives (New York 1916)

Works by Hiram Percy Maxim
Life's Place in the Cosmos (New York and London 1933)
A Genius in the Family: Sir Hiram Maxim through a Small Son's Eyes (Michael Joseph 1936, repr. Dover Publications 1962)
Horseless Carriage Days (New York and London 1937)

Other books
Allfrey, Anthony, *Man of Arms: the life and legend of Sir Basil Zaharoff* (Weidenfeld and Nicolson 1989)
Ford, Roger, *The Grim Reaper: the Machine-gun and Machine-gunners* (Sidgwick and Jackson 1996, repr. London 1997)
Gibbs-Smith, C H, *Aviation: a historical survey from its origins to the end of World War II* (London 1970)
Goldsmith, Dolf L, *The Devil's Paintbrush, Sir Hiram Maxim's Gun* (Toronto 1989)
Gollin, Alfred, *No Longer an Island: Britain and the Wright Brothers, 1902-09* (London 1984)
Hamilton, James E, *The 'Chronic Inventor': the Life and Work of Hiram Stevens Maxim* (London Borough of Bexley Libraries and Museums Department 1991)

Harrison, Tony, *Square Rounds* (London 1992)

Hartcup, Guy, *The War of Invention: Scientific Developments 1914-18* (London 1988)

Jarrett, Philip, *Another Icarus: Percy Pilcher and the Quest for Flight* (Washington DC 1987)

Johnson, Clifton (ed), *Hudson Maxim, Reminiscences and Comments* (London 1924: repr by New York 1927 as *The Rise of an American Inventor: Hudson Maxim's Life Story*)

Lee, Percy Maxim and Glessner, John, *Family Reunion, an incomplete account of the Maxim-Lee family history* (privately published 1971)

Moolman, Valerie, *The Road to Kitty Hawk* (Time-Life Books 1980)

Mottelay, P Fleury, *The Life and Work of Sir Hiram Maxim* (London 1920)

Penrose, Harald, *British Aviation, the Pioneer Years, 1903-14* (Rev. ed. London 1980)

Reader, W J, *Imperial Chemical Industries, a History: Vol I: The Forerunners, 1870-1926* (Oxford 1970)

Scott, J D, *Vickers, a History* (Weidenfeld and Nicolson 1962)

Terraine, John, *White Heat: The New Warfare 1914-18* (London 1982)

Trebilcock, Clive, *The Vickers Brothers, Armaments and Enterprise, 1854-1914* (London 1977)

Index

Abbreviations
FRS = Fellow of the Royal Society; RA = Royal Artillery; RE = Royal Engineers; RN = Royal Navy; US = United States; USN = United States Navy; VC = Victoria Cross

Abbot, Maine 17, 19
Abel, Sir Frederick 67, 72, 93, 143
Ader, Clément 96, 103, 107, 112, 118fn, 120
Aerial League of the British Empire 175
Aeronautical Society of Great Britain 97, 100, 105, 110, 115, 117, 163, 171, 173
Alger, Professor P R, USN 83, 126, 152, 214
Amalgamated Society of Engineers 69
American Civil War 19, 47, 57, 63, 95, 148
American Engineer and Railroad Journal 108
Anderson, Sir William, 125fn 127
Argonauts of the Air, The 110
Argyll, Duke of 175
Armstrong, Sir William (later Lord Armstrong) 42, 50
Armstrong, Sir W G, and Company (later Armstrong, Whitworth) 62, 79, 93-4, 125, 129, 146, 166
Army Balloon Factory, Aldershot 110
Artificial and Natural Flight 96, 119, 170
Atlantic City pageant 210
Autocar 119

Baden-Powell, B F S 109, 115, 163
Battle Cry of Peace, The 197
Beecher, Henry Ward 24, 27
Belloc, Hilaire 186
Beresford, Lord Charles 55, 60
Bernadou, John B, USN 150
Bexley Cottage Hospital, demonstration of flying machine at 112
Birmingham Small Arms Company 47
Black Tom explosion, New York, 1916 206-7
Blériot, Louis 173, 176
Bliss E W and Company of Brooklyn 153
Bloch, I S 192-3
Boer War 146, 149, 160
Brasher's Falls, New York State 21
Brennan, Louis, and torpedo 83fn
Brewer, Griffith, balloonist 104fn
British Expeditionary Force 187, 194
Browning, John M 70
Browning machine gun *see* Colt Firearms Company
Budden, Jane *see* Maxim, Jane Budden

Callwell, Captain C E, RA 144
Cambridge, Duke of 48, 54, 67, 79, 92

Capper, Colonel John 169, 175
captive flying machine 164-5
Carnegie, Andrew 168
Cassel, Ernest, financier 63, 73
Cayley, Sir George 97
Channel tunnel 184
Chanute, Octave 101, 103, 105, 108-9, 111, 116, 118, 139
Churchill, Rt Hon Winston S 171-2fn, 175
Clarke, Sir Andrew, RE 57, 59
Cleveland, Grover, US president 39fn
Cloete, W Brodrick 52fn, 70, 74, 76, 91, 96-7, 110, 114, 143
Cody, Samuel Franklin 163-5, 170, 173-4
coffee concentrate 25, 187
Colonial Motion Picture Corporation 190
Colt Firearms Company, Hartford, Connecticut 70, 86, 145, 186, 200, 204
Columbia Powder Company, New Jersey 75-7, 83
Commerell, Admiral Sir Edmund, VC 81, 105, 113
Committee of Imperial Defence 172
Crookes, Sir William, FRS 7
Curtiss, Glenn 176fn

Dawson, Lieutenant A Trevor, RN 160, 178
Dee, Michael 214
Defenseless America 196-7, 201
Deutsche Waffen und Munitionsfabriken (DWM) 81fn, 146-8, 166, 186
Dewar, Sir James 67, 72, 93, 143
Dewitz, Baron Hrolf von 196
Dexter, Maine 18, 20
Doewe, Heinrich, and his cuirass 92
Douglas, Bryce 56
Drake, Oliver P 21
Du Pont de Nemours company, E I interest in smokeless propellant 73, takes out option on Hudson's explosives 87, exercises option 125, collaborates with Hudson during Spanish-American War 131-6, manufactures his explosives 177, and delayed-action fuse 178, renews agreement with Hudson 188, distributes his *Defenseless America* 197, profits during First World War 202, commiserates on Hudson's death 213.

du Pont, Alfred 73
du Pont, Eugene 154
du Pont, Francis I 188
du Pont, Pierre S 155, 188, 196, 198
Dulwich College, Kent 51, 53
Dunne, John W 165, 166fn
Durban, Lilian *see* Maxim, Lilian Durban
Durban, the Rev Dr William 88-9, 143, 159, 176-7, 190
Durban, Will 129, 132, 138, 215
Dynamite Stories 159

Edison, Thomas Alva 28-9, 31, 42, 45, 72, 195, 201
Edison & Swan Electrical Company 54
Edward, Prince of Wales (later Edward VII) 54, 60, 145-6
Elswick Ordnance Company, Newcastle 50, 129
Engineer, The 53fn, 61, 99, 104, 126, 130, 204
Engineering 31-2, 42, 69, 105-6, 108, 133-4, 180-1, 183
Enricht process 200
Esnault-Pelterie, R, and REP monoplane 175-6, 179
Eynsford firing range, Kent 82, 115, 117, 123

Farnborough, Balloon and Army Aircraft Factory 166, 169, 170, 172
Fitchburg, Massachusetts 21-2, 45
Flirt, steam launch 30
Flynt, Daniel 17
Forcite Powder Works, Lake Hopatcong, New Jersey 154
Ford, Henry 201
Franco-Prussian War 41, 45, 48, 61, 95
Fung Ling, Chinese Imperial Legation 82-3
fuse, delayed-action 178

Gardner, William, and Gardner gun 47-8, 53, 56
gas, apparatus for dissipating poison 203
Gathmann, Louis, and his high explosive shell 152
Gatling, Richard J, and Gatling gun 45, 47-8, 50, 53, 55, 69
George, Duke of Kent (later George V) 113-4
Georgia, US battleship 181-2
Gibbs-Smith, C H 119, 120
Giffard, Henri, balloonist 95
Graham, Lieut-General Sir Gerald, VC 63, 81
Grahame-White, Claude, aviator 120, 176-7
Griffin, William, mechanic 65-6
Guthrie, Arthur, mechanic 103, 107

Haber, Fritz 8
Hague Peace Conferences 168, 193
Haldane R B 169, 172

Hamilton, Jack ('Captain Graystone') 44, 46
Hamilton, Josephine (Mrs Hiram Percy Maxim) 139, 140, 206
Hargrave, Lawrence 116, 118
Hawk glider 115-8
Haynes, Mr and Mrs Charles, of Boston 35-6, 76
Haynes, Sarah *see* Maxim, Sarah Haynes
helicopter 17, 117
Hendon aerodrome 176
Henson W S 97
Hewitt, Edward R 31, 98, 120
Hoosac Tunnel 72
Hotchkiss, Benjamin B, and Hotchkiss gun 45, 48, 53, 79, 93, 167, 185

incendiary bullets 203
Indian Head, Maryland 134, 151, 178
International Electricity Exhibitions, Paris 28, 31, 41
International Exhibition of Motors and their Appliances, Imperial Institute, South Kensington 115
International Inventions Exhibition, Kensington 53, 61

Jackson, Thomas, mechanic 103, 107
Jéna, French battleship 180, 184
Johnson, Clifton 9, 197, 207-9, 213, 215
Joubert, Eldon 167
Joubert, Maxim 167, 205, 215
Jutland, battle of 184

Kabath, Nicholas de 32, 67, 80-1
Kelvin, Lord 110, 114, 118
Kent's Hill Maine Wesleyan Seminary 24, 67
Kitchener, General Sir Herbert (later Field-Marshal Lord) 144, 203
Kitty Hawk, North Carolina 118
Knowles, Alden W 33, 64
Krupp company of Essen 61, 147, 201
Krupp, Friedrich 60

Lanchester, F W 87, 119
Langley, Samuel Pierpont 76, 96, 99, 101-3, 109, 116-8, 120, 163
Le Fleming, Vera 178
Leavitt, Frank M 153, 155, 214
Lee, Percy Maxim 9, 10, 206, 215
Lehigh Valley railway company 206
Leighton, Helen ('Nell Malcolm') 36, 43, 52, 76, 138-141
Li Hung Chang, Chinese statesman 82, 187
Liberté, French battleship 184
Liddell Hart, Sir Basil 7
Lilienthal, Otto 111, 114-5, 117
Lloyd George, David 194, 215
Loewe, Ludwig 80, 81
Loewe, Sigmund 81, 83, 91-4, 106, 117, 123-4, 144, 148, 160

Lusitania, sinking of 196

'Maxicream' and 'Maximfeast' 206
machine gun, Maxim automatic
    inception 44, early development of 46-9,
    formation of Maxim Gun Company 52,
    overseas sales drive 59, use in colonial
    wars 79, at Omdurman 144, and in Boer
    War 146, manufacture in Germany and
    Russia 186, effect on tactics during the
    First World War 193-4
MacRoberts, works manager, Nobel company
    57, 75
Maine, US battleship 131fn
Marshall, Frank J, chess champion 195
Massachusetts Institute of Technology 52,
    71
Maxim & Welch, Gas and Steam Engineers,
    New York 25
Maxim (formerly Squankum), New Jersey 75,
    77 83, 86
Maxim Aerial Torpedo, The 126, 128
Maxim Auto-Car Syndicate, London 94, 123
Maxim Carbide and Acetyline Gas Syndicate,
    London 94, 123
Maxim Electrical and Engineering Company,
    Pimlico, London 54, 165
Maxim Gas Machine Company, New York
    23
Maxim Gun Company, London 52, 65
Maxim Munitions Corporation, New York
    198-9, 200-201, 210, 214
Maxim Nordenfelt Guns and Ammunition
    Company
    London formation 63, initial difficulties 68,
    dealings with Hudson Maxim 73 *et seq*,
    amalgamation with Vickers, Sons and
    Maxim 94
Maxim Park, Lake Hopatcong 157
Maxim Powder and Torpedo Company, New
    York and London 83, 86-7, 90-1, 123, 126
Maxim Silencer Company 199
Maxim, Adelaide ('Addie') 27, 50, 167
Maxim, Eliza 16, 24
Maxim, Florence 22, 27, 50, 139
Maxim, Frank 16, 28
Maxim, Harriet Stevens 14, 23, 35, 38, 76,
    90, 139, 142, 148-9
Maxim, Henry 15, 19
Maxim, Hiram Percy
    born 22, early life 26 *et seq*, repudiated by
    Hiram 52, designs automobiles for Pope
    Manufacturing Company 139, involved in
    father's trial for bigamy 140-1, marries
    Josephine Hamilton 141, represents
    Hiram at Sangerville centenary 190,
    Maxim Silencer Company, 199
Maxim, (Sir) Hiram Stevens
    early life 16 *et seq*, travels during the Civil
    War 20, work on gas machines 21,

marries Jane Budden 22, work on electric
    lighting apparatus 28, involvement with
    Helen Leighton 36, and with Sarah
    Haynes 37, settles in Paris 39, works on
    automatic gun 45 *et seq*, and smokeless
    propellant 57, settles in England 53,
    marries Sarah Haynes 62, work on flying
    machine 96 *et seq*, quarrels with Hudson
    136 *et seq*, trial for bigamy 139, receives
    knighthood 145, intervention during Boer
    War 149, his views on Casino at Monte
    Carlo 162, venture with captive flying
    machine 164, modified aeroplane of 1910
    174, resigns from Vickers 176,
    controversy with American authorities
    over smokeless propellant 180 *et seq*,
    activities during the First World War 202
    et seq, death 204
Maxim, Hudson Day 77, 90, 189, 190, 215
Maxim, Hudson
    early life 17, assists Hiram in New York
    23, enrols at Kent's Hill Wesleyan
    Seminary 24, establishes publishing
    business 34, marries Jane Morrow 65,
    joins Hiram in England 65, work on
    smokeless propellants 72, loses hand 84,
    marries Lilian Durban 89, his company
    taken over by Du Ponts 125, 132, quarrels
    with Hiram 132 *et seq*, profits from
    Spanish-American War 152, work on
    motorite propellant 154, develops
    explosives and other war material for US
    army and navy 156, 177, buys land at
    Lake Hopatcong 156, dispute with Morris
    Canal company 189, joins Civilian
    Advisory Board 195, active in
    preparedness movement 195 *et seq*,
    involvement with Maxim Munitions
    Corporation 198, collaboration with
    Clifton Johnson 207 *et seq*, death 213
Maxim, Isaac Weston 13-16, 23, 35, 38, 45
Maxim, Jane Budden 22, 26, 30, 37, 50-2, 61,
    139
Maxim, Jennie Morrow 64-5, 77, 128, 189
Maxim, Leander 16, 19
Maxim, Lilian Durban
    courted by and marriage to Hudson 88-9,
    travels to the United States 129, involved
    in property dealings at Lake Hopatcong
    157, eventful social life 157, reaction to
    Hudson Day Maxim 190, encourages
    writing of *Reminiscences and Comments* 207,
    trip to England (1925) 211, marriage to
    Michael Dee 214, death 215
Maxim, Lucy 15, 20, 24
Maxim, Samuel
    born 16, lack of ambition 28, gives reading
    at Maxim family gathering 35,
    photographed with Hiram and Sarah
    (1891) 76, attempts to mediate in quarrel

between Hiram and Hudson 137, poverty and ill-health 212, death 215

Maxim, (Lady) Sarah Haynes 36, 39, 43, 47, 51
marriage to Hiram 62, accompanies him on visits to Maine (1891) 76, and the Riviera 81, their marital relationship 128, involvement in bigamy trial 140, assists Hiram in writing *My Life* 202, later contacts with Hudson and Lilian 211, corresponds with Lilian Durban 213, death 215

Maxim-Schupphaus smokeless powder 84-6, 91, 93, 123, 125-7 *et seq*, 133, 143, 150-1, 180, 202

Maxim-Weston Electric Light and Power Generating Company, London 31, 43, 46

Maximite (Hudson's high explosive) 126-7, 132, 134, 136, 150-3, 178, 202

Maximite (Hiram's smokeless powder) 71, 73-4, 93

Mayfly, rigid naval airship 173, 179

Mikasa, Japanese battleship 184

Miller, William, and Millerite movement 15

Monte Carlo, Casino at 81, 162

Montgolfier brothers 95

Montreal, Canada 20

Morris Canal project 189, 214

mosquitos, strange behaviour of 29-30

'motorite' 154-5

Moulton, J Fletcher (later Lord Moulton) 86, 96, 105fn, 125, 202, 205

mousetrap, automatic 17

Mowbray, Professor G M 72-5, 77, 214

Munroe, Professor Charles E 135, 150

Naval Construction and Armaments Company, Barrow 94fn

Nobel (Anglo-German) Dynamite Trust 77

Nobel, Alfred 58, 66, 67, 70-1, 86, 93

Noble, Sir Andrew 94, 129, 130-1, 168, 181

Nordenfelt, Thorsten, and his gun 47-8, 53, 55, 58-9, 60-3, 67-9, 79, 82

Novelty Ironworks and Shipbuilding Company, New York 22

Nulli Secundus, military airship 170

Oberhardt, William, artist 211

Omdurman, battle of 144, 147

Orneville, Maine 16, 17

Palmer, Johnny 44, 46

Penobscot Indians 16

Penrose, Harald 119

Philadelphia Centennial International Exhibition (1876) 27

Phillips, Horatio 98

Picatinny Arsenal, New Jersey 154, 177

Pilcher, Percy Sinclair 114-8

Piscataquis County, Maine 14 and *Piscataquis Observer* 18

Pittsfield, Massachusetts 34, 63, 67, 70, 72, 74, 77

Puckle, James, and his gun 8

Rayleigh, Lord 105, 110

Rodman, Thomas J 57, 134

Romaine (Dennison) 36, 43, 76, 141, 205

Roosevelt, Theodore, (later US president) 131, 133

Rosslyn, Lord 162, 163fn

Rothschild, 1st Baron 50, 60, 63, 80-1, 94, 193

Royal Arsenal, Woolwich 62, 67, 146

Royal Flying Corps 179, 203fn

Royal Small Arms Factory, Enfield 63, 69, 79, 204

Royal United Services Institution 45, 55, 126

Russo-Japanese war 166, 186

Salisbury, Lord 160

Sandy Hook, firing ranges at 70-1, 74, 85-6, 127, 134

Sangerville, Maine 14, 17, 76, 190-1

School of Musketry, Hythe 68, 144

Schupphaus, Dr Robert 77-8, 83-6, 125-6, 156

Schuyler, Spencer D 29, 30, 71, 73-4, 83, 86

Science of Poetry and Philosophy of Language, The 159, 160

Scientific American 70, 76, 85, 136, 143, 151, 157

Scottish Aeronautical Society 188

Shields G R 79, 91-2

Silverman, Louis, foreman mechanic 55, 59, 80

Sims, F R, automobile engineer 169fn

Smith, Bucknall of *The Strand Magazine* 104, 112-3

Smithsonian Institution, Washington 76, 99

smokeless powder, controversy over (1910-11) 179-183

Société des Munitions Françaises 166

Society of the Chemical Industry 126

Somme, battle of the 203-4

soya bean 206

Spanish-American war 118, 131, 153

Sperry, William 154-5

Spotsylvania Courthouse, battle of 19

Springall, Dr 20

Stanley, Henry M, explorer 59

Stevens, uncle Amos 19, 43

Stevens, uncle Levi 21-3, 45

Stone, F G, RA 61, 78, 173

Strand Magazine, The 44fn, 85, 112

Stringfellow, John 97

Sutherland, Duke of, (George Leveson-Gower) 54, 58, 100

Sweat, Daniel 16

Symon, Robert 50, 52, 56, 76, 96-7, 143

Taft, William H, US president 181
Thurston, Dr A P 101fn, 164fn, 173fn, 174
Titanic, and Sir Hiram Maxim's anti-collision
  device 184-5
Turpin, Eugène, explosives scientist 71fn

Union Metallic Cartridge Company,
  Bridgeport, Connecticut 29, 36
US Bureau of Ordnance 70, 73-4, 83-6, 124,
  131, 150-1, 178, 183, 200
US Civilian Advisory Board (later Naval
  Consulting Board) 195, 201
US Department of the Navy 73-4, 84, 129,
  131, 150-1, 153, 155, 178
US Electric Lighting Company, New York
  29, 31, 41-2, 63
US Naval Torpedo Station, Newport, Rhode
  Island 135, 150, 155
Upper Austin Lodge, Eynsford, Kent 115

Vesuvius, US dynamite cruiser 71, 136
Vickers, Albert 50, 52, 55, 59, 62, 97, 103,
  112, 204, 215
Vickers machine gun 187, 198, 204
Vickers, Sons and Maxim 94, 117, 122, 130,
  132, 143, 146, 161, 166, 169
Vieille, Paul, explosives scientist 66, 71

Vos Inhalatorium, Nice 161-3

Washington Navy Yard 69, 134, 178
Wayne, Maine 25, 35, 38, 137, 148
Webster, Sir Richard 86
Wells, H G 105, 110, 111, 165
Wenham F H 98fn
Whitehead automobile torpedo 56, 127, 153,
  155
Wilmington range, Dartford 78
Winchester Repeating Arms Company 45,
  64-5, 73
Winthrop, Maine 13, 14, 24, 35
Wolseley Tool and Motor Car Company
  160, 163, 169
Wolseley, Sir Garnet (later Field-Marshal
  Lord) 54, 58, 67, 79
Wright brothers, Orville and Wilbur 104,
  109, 117-9, 169, 170, 174

Xylonite explosives company 72, 77

York, George, Duke of (later George V) 113

Zaharoff, Basil 59, 60, 62, 67, 74, 80, 81, 91,
  94, 105, 111, 129, 137, 148, 166–7, 215
Zalinski, E L, and his dynamic gun 71, 90,
  136fn
Zeppelin, Count, and airship 163, 172, 203